An Introduction to Parasitology

The Institute of Biology aims to advance both the science and practice of biology. Besides providing the general editors for this series, the Institute publishes two journals *Biologist* and the *Journal of Biological Education*, conducts examinations, arranges national and local meetings and represents the views of its members to government and other bodies. The emphasis of the *Studies in Biology* will be on subjects covering major parts of first-year undergraduate courses. We will be publishing new editions of the 'bestsellers' as well as publishing additional new titles.

Titles available in this series

An Introduction to Genetic Engineering, D. S. T. Nicholl

Photosynthesis, 5th edition. D. O. Hall and K. K. Rao

Introducing Microbiology, J. Heritage, E. G. V. Evans and R. A. Killington

Biotechnology, 3rd edition, J. E. Smith

An Introduction to Parasitology, Bernard E. Matthews

This introductory undergraduate textbook provides a concise, clear and affordable overview of parasite biology for students and non-specialists, to equip them for approaching more technical and detailed literature. Using examples from all groups of animal parasites, the text considers the various stages of a parasite's life cycle from finding hosts to surviving within the host. The significance of damage caused by parasites and approaches to control are discussed, giving the book a well-rounded approach for those new to the subject or revising key points. A key feature of the book is its comparative rather than systematic approach. Suitable for all introductory parasitology courses in biology, zoology and human and veterinary medicine.

An Introduction to Parasitology

Bernard E. Matthews

Institute of Cell, Animal and Population Biology,
University of Edinburgh

CAMBRIDGE
UNIVERSITY PRESS

PUBLISHED BY THE PRESS SYNDICATE OF THE UNIVERSITY OF CAMBRIDGE
The Pitt Building, Trumpington Street, Cambridge CB2 1RP, United Kingdom

CAMBRIDGE UNIVERSITY PRESS
The Edinburgh Building, Cambridge CB2 2RU, United Kingdom
40 West 20th Street, New York, NY 10011-4211, USA
10 Stamford Road, Oakleigh, Melbourne 3166, Australia

First published 1998

Printed in the United Kingdom at the University Press, Cambridge

Typeset in Monotype Garamond 11/13   [SE]

A catalogue record for this book is available from the British Library

ISBN 0 521 57170 7 hardback
ISBN 0 521 57691 1 paperback

Contents

Preface

Modern parasitology embraces an enormous range of biological disciplines from the molecular to the global, and this is reflected in its literature. In general, parasitological books fall into two major categories. There are the large, comprehensive textbooks that aim to cover the whole subject and there are specialised volumes that consider a single group, species or disease in considerable depth, providing up-to-date, definitive information for the specialist. Examples of both are included in the reading list at the end of this book.

The present volume does not attempt to emulate either of these, rather it offers a short, hopefully comprehensible, introduction to aspects of parasitology which will equip the reader with the concepts and vocabulary needed to appreciate the more specialised literature. Towards this end technical terms have been emboldened in the text and defined in a glossary at the end.

The biggest difficulty has been deciding what to include. The choice of the material covered has had to be a personal one and inevitably many topics have been omitted. The danger of any such selection process is that one will succeed in satisfying none of one's readership. I hope this is not the case and I extend my apologies to those of my colleagues whose favourite parasites are not included or whose work has apparently been ignored. No slight was intended and space was at a premium.

My thanks are due to the colleagues who have generously allowed me to use their photographs and with whom I have discussed parts of the text, but all errors are of course mine. Finally I must thank my family for their tolerance during the preparation of the manuscript and I dedicate it to them.

Bernard E. Matthews   Edinburgh, 1997

1

The spectrum of animal–animal associations

Intra-specific associations

With the possible exception of the castaway sitting with his eight records on that mythical desert island, we all rely to a greater or lesser extent on other people. We have associations with them. These personal relationships vary from the remote, the people who provide services that we only notice when they are withdrawn because of industrial (in)action; through the casual, the person we try to ignore on the bus; to the intimate, our immediate family and close friends.

Similar associations occur between animals of the same species (**intra-specific associations**). The range of these associations is enormous and attempts to categorise them are fraught with difficulty as there are so many exceptions. It is, however, possible to offer a number of illustrative examples. These are presented in order of increasing permanence and dependence between the partners.

Herd/flock associations

A herd or flock is generally a loosely knit community which individual animals are free to join or leave. While the size of the herd makes it conspicuous and may attract the attention of predators it is generally the sick and young animals that are vulnerable. Increased numbers provide protection, as some individuals can be on watch while others feed, so that there is continuous vigilance for the group as a whole and greater protection for the majority. When predators

Table 1.1. *Relationship between female fecundity and parental care in a number of teleost fish species*

	Number of eggs in batch	Parental care
Cod (*Gadus morhua*)	4,000,000	None
Herring (*Clupea harengus*)	35,000	None
Gunnel, Butterfish (*Pholis gunnellus*)	400	Parents guard egg mass on shore
Siamese fighting fish (*Betta splendens*)	150	Male guards bubble nest
Stickleback (*Gasterosteus aculeatus*)	75	Male guards nest
Sea horse (*Hippocampus*)	55	Male brood pouch
Tilapia	35	Female broods in mouth

do strike they tend to attack individual, outlying animals and there is often a movement of individuals towards the centre of the group where protection is greatest. Interestingly this protection is not only from macropredators but also from biting flies and other **ectoparasites**. Individual animals in larger groups have fewer flies on them and those in the centre of the group have fewer than those at the periphery. Herds, flocks, shoals and swarms also provide advantages in mate location and breeding, indeed some species only aggregate during the breeding season. These associations are thus generally beneficial but not obligatory and many species are largely solitary except for short periods of mating.

Parental care

The amount of effort devoted to parental care by different species varies greatly and seems to be inversely proportional to the number of offspring produced (Table 1.1).

At one extreme are those animals that shed their gametes into the environment and take no further interest in their development or well-being. Seasonality, shoaling, aggregation and communal habits all tend to encourage the union of gametes, but the wastage of zygotes is enormous. In human terms this may appear profligate and irresponsible, but the larval stages of, for example, marine invertebrates provide much of the animal plankton that is so important to marine food webs.

At the other extreme are the higher mammals with internal fertilisation and development and the birth of a small number of relatively well developed offspring. In the case of the primates birth may be followed by several years of more or less continuous attention until the young are fully capable of leading an independent existence. In some cases, e.g. elephants, care of the young is shared by different members of a larger family group so that the biological parents do not have sole responsibility for raising their young. This is, of course, another advantage of living in a group of the same species.

Between these two extremes are many intermediate stages. There are many examples of marine annelids, crustaceans, molluscs, fish and other species that brood their eggs in burrows or nests or about their bodies. Parental care is not necessarily the responsibility of both parents or even of the mother. It is male Siamese fighting fish (*Betta splendens*) and three-spined sticklebacks (*Gasterosteus aculeatus*) that build nests, entice their females to them and then guard the fertilised eggs until they hatch. Male sea horses (*Hippocampus*) have a brood pouch beneath the tail and eggs are transferred to this during mating. Here they remain until they have hatched and the fry are able to fend for themselves. In the case of the midwife toad (*Pipa*) it is the female that provides the brooding protection. The eggs are not laid in water as they are for other amphibians, but are deposited under flaps of skin on the back of the female. Here they hatch and metamorphose so that small toads emerge from their mothers. Once the young have left the parent there is often little effort to provide long-term protection, and indeed in some cases an unwary offspring may provide a tasty mouthful for a hungry parent!

Birds, of course, take greater care of their progeny and will feed and care for them until they are relatively well developed, able to fly and to feed independently. This, in the case of some of the larger species, may take a year or more and thus reduces the reproductive potential of the parents. A balance is thus achieved between the number of progeny and the effort that the parents put into raising their young.

Pairing

The reproductive imperative is strong in all species but the outward signs of intimacy vary enormously. **Hermaphrodite** organisms which contain both male and female organs may appear to have overcome the difficulties of locating a suitable mate, but for many of these species cross-fertilisation is the rule so they face similar problems to **dioecious** species.

Many species show no appreciation of the opposite sex and gamete production and release appear to be governed more by environmental factors than by association. Other species go to considerable lengths to attract a mate even though for most of the time they are solitary. The **pheromone** attractants produced by some female moths are legendary, being able to attract males from several kilometres away. Yet other species only come into breeding condition when in more or less permanent relationship with a mate, and mate attraction becomes even more important for these. In many cases, once the partners have paired and mated the association is terminated and many males go off to seek other females, but for other species the pair may remain together for a time after mating. If parental care is to be provided by both parents then the relationship between them must last for at least the time needed to raise their young.

Some species, e.g. swans, are **monogamous** and mate for life, but each member of the pair is capable of an independent existence. There are species where one of the members, usually the male, gives up its independence to remain permanently with the female. Thus the diminutive males of some deep sea angler fish remain permanently attached to their much larger partners and share a blood supply. It is assumed that this unusual mode of life is in response to the extreme environments of the fish and the difficulties of locating a mate in the abyssal depths. Probably the most extreme example of togetherness is illustrated by *Trichosomoides crassicauda,* a nematode parasite of the bladder of rodents in which the male spends his life within the uterus of the female. It is difficult to conceive of a more intimate association.

Colonial life

Not all colonial animals exhibit co-operation between individuals, indeed there may be competition between individuals, especially among juvenile stages for advantageous sites. Thus barnacles and mussels live in aggregations of individuals brought together by ecological and environmental factors and are probably better considered more like herds than true colonies. However, the most intimate and interdependent of intra-specific relationships are exhibited by animals that live colonially. The interrelationships between morphologically and physiologically different individuals with consequent division of labour in an ant or bee colony are well known. In some cases the colony may behave in such a co-ordinated way that it acts virtually as a single enormous individual. Even amongst colonial animals there is a gradation in levels of intimacy and interdependence. A single worker bee may be considered as an individual (it can sting you) as well as part of the greater whole, the colony. This is not true for the constituent polyps of the colonial coelenterates (Siphonophora). The Portuguese man-of-war (*Physalia*) and by-the-wind-sailor (*Velella*) drift on the ocean currents and may be seen on the shores of Britain during the summer. While these appear to be and are often considered individual organisms they in fact consist of a number of polyps, each with a specialised function for buoyancy, feeding, defence, reproduction, etc., much as the worker, soldier and queen ants have their own specialised functions within an ant colony. The only real difference is that the individual coelenterate polyps are no longer capable of individual locomotion.

Intriguing as these intra-specific associations are they are more the province of the animal behaviourist than the parasitologist, and the remainder of this text is concerned with associations between animals of different species (**inter-specific associations**).

Inter-specific associations

Emerson (1803–1882) once said '*He shall be as a god to me, who can rightly divide and define*', and there have been many attempts to achieve such division and definition of the vast range of inter-specific associations that have evolved. None of these attempts has been wholly successful and different authors have provided terminology to suit their own purposes. The present author is no exception and some definitions will be offered below. However, he is making no claims for deification and is well aware of the inadequacies of some of the definitions offered. Evolution has provided an infinity of variations around the theme. One concept that may be helpful is to consider the range of relationships not as a series of discretely defined entities but more as a continuum, with one type of association merging into the next rather in the way that the individual colours of the light spectrum merge almost imperceptibly from one to the next.

The extremes of such a sequence are relatively easy to recognise. No one would question that a hunting lion is a predator and the wildebeest its prey, or, on a different scale, that the spider is the predator and the fly trapped in its web the prey. The situation becomes more complicated with different relative sizes of predator and prey. A spider may trap an insect much larger than itself, stun it and over a period of time remove its body fluids as food. The prey is

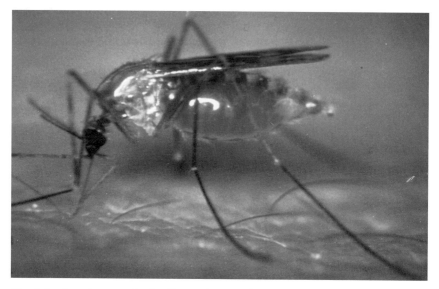

Fig. 1.1 Female mosquito feeding on human skin. Depending on context she may be regarded as a micropredator or an ectoparasite. In addition she is likely to be a vector capable of carrying a range of disease-causing organisms from one host to another.

killed but not immediately, and the relationship is still recognisably predator/prey. But what if the insect had not been killed? This situation occurs when a mosquito takes a blood meal from a human (Fig. 1.1) or a leech attaches to a fish to suck its blood. The 'prey' item is not killed and, in many cases, except for a certain amount of nuisance, suffers little or no appreciable harm. The principle of attacking a larger animal and removing some of its body fluid is common to the spider, the mosquito and the leech, and some authors have used the term **micropredator** to cover all of these situations. Some 'micropredators', e.g. ticks and mites, may spend an appreciable proportion of their life on another animal and, because of their reproductive potential and blood feeding behaviour, cause extensive direct damage. Others, e.g. mosquitoes and Tsetse flies, may transmit disease-causing organisms. Many micropredators are often referred to as **ectoparasites**, (literally *ecto* – outside, *para* – beside, *sitos* – food) but again the principle of removing some of the host's body fluids is the same in each case. To add further to the confusion there are species, usually but not exclusively insects, which spend their larval stages within another species, again usually an insect, that they kill when they emerge as adults. These **parasitoids** have some of the attributes of a predator in that they eventually kill their host and some of the features of a parasite as they live part of their lives within it without killing it. Thus, even setting the first point in our spectrum is not as simple as might at first appear.

At the opposite end of the range are the mutualists. These are animals that are so interdependent that neither of the partners is able to maintain an independent existence, and removal of one results in the death of the other. Between the two limits of predator/prey relationships on the one hand and mutualists on the other there is a vast array of more or less intimate and more or less obligatory inter-specific relationships. The general term **symbiosis** (literally *sym* – together, *bios* – life) is often used to cover all categories of intimate inter-specific associations, but even here a word of caution is needed as some authors use the term symbiosis to cover a much more restricted set of non-obligatory, mutually beneficial associations.

Under the umbrella term symbiosis it is possible to recognise a number of sub-divisions that may be more or less clearly defined. In order of increasing intimacy and dependence these are commensalism, parasitism and mutualism, but as we will see there may be further sub-divisions within these.

Commensalism

Commensalism (literally *co* – together, *mensa* – table) is the least intimate and least obligatory of the generally recognised inter-specific associations. Both partners are able to lead independent lives but one or both may gain advantage from the association when they are together. Commensalism is widespread throughout the animal kingdom and a number of classes of commensalism have been recognised based on the biological nature of the bond between the partners. It should be noted that the associations recognised in the different divisions of commensalism are not exclusive to inter-specific associates; cleaning, protection, transport, etc. may be practised by partners in intra-specific relationships as well.

Cleaning commensalism Possibly the best-known and best-studied examples of commensalism come from the relationships between larger fish and small fish and shrimps which remove fungi and other ectoparasites from their bodies. The large fish encourage feeding by approaching specific, prominent 'cleaning sites' on the reef and taking up characteristic postures that indicate their non-aggressive intent. The cleaners then approach, remove and eat offending ectoparasites. This cleaning relationship has recently been investigated as a possible control measure against ectoparasitic sea lice on farmed Atlantic salmon kept in sea cages off Scotland and Norway. Initial experiments introducing wrasse (*Ctenolabrus* and *Labrus*) into the cages have been encouraging and may provide an environmentally friendly way of controlling a major problem.

Cleaning commensalism is not restricted to aquatic environments. There are, among others land crabs that remove ectoparasitic ticks from sun bathing Galapagos iguanas; Egyptian plovers (*Pluvianus aegyptius*), which were known to Herodatus (*c.* 480–*c.* 425 BC) as crocodile birds because of their habit of entering the mouths of Nile crocodiles to feed; and cattle egrets (*Bubulcus ibis*) which remove ectoparasites from a wide range of African mammals. Red billed oxpeckers (*Buphagus erythrorhynchus*) spend much of their lives on large African savannah mammals feeding on ectoparasitic arthropods, especially ticks. They may also give their partners advance warning of approaching predators so that both partners gain advantage from the association. However, the ticks that are removed contain some blood. The birds may gain a taste for blood to such an extent that they take it directly from unhealed lesions in the skin. They do not directly damage intact skin themselves but do delay the healing of existing wounds and may increase the chance of secondary bacterial infection. The balance of the relationship has thus been tilted slightly in favour of one of the partners, the oxpecker, to the detriment of the other.

The kea (*Nestor notabilis*) is a New Zealand parrot. During the summer it feeds mainly on fruit and insects but in the winter when such food is scarce it moves on to sheep and feeds on ectoparasites. When these run out it will attack the sheep directly, clinging to the wool near the tail and digging a deep hole with its powerful beak to feed on the fat around the kidneys. The relationship between the sheep and the kea thus changes during the year and, from being mutually beneficial becomes something much more sinister. This again illustrates the difficulty in presenting a rigorous definition for any association: the relationship may change as circumstances change.

Protective commensalism   The hermit crab (*Eupagurus bernhardus*) lives in the discarded shells of the edible whelk (*Buccinum undatum*). Often the whelk shell has one or more sea anemones (*Calliactis parasitica*) attached to it. The crab transports the otherwise sessile anemones and it is probable that the anemones also gain extra food particles from the feeding of the crab. Whether the crab gains anything from the association is problematical. However the Mediterranean hermit crab (*Dardanus arrosor*), which lives in rather more fragile shells, actively transfers anemones on to its shell, and it has been shown that crabs with anemones are less liable to predation from octopus. Thus both partners benefit from their association but are able to live independently.

Fig. 1.2 Protective commensalism. The cloak anemone, *Adamsia palliata*, on a hermit crab, *Eupagurus prideauxi*. It is doubtful if the anemone provides much protection to the crab and this may be more an example of phoresis, but other crab/anemone associations have been shown to provide positive protective benefits.

The anemone (*Adamsia palliata*) which lives around shells occupied by the smaller hermit crab (*Eupagurus prideauxi*) (Fig. 1.2) is more commonly found on British shores. Normally when a hermit crab grows it moves to a larger shell, but *Adamsia* secretes a horny substance that effectively extends the shell making a move unnecessary. The anemone has a vested interest in maintaining the relationship as the crab is able to seek a new shell but a discarded anemone dies. This association, once established, is thus obligatory on one side but not the other.

Transport commensalism (phoresis) Transport commensalism or **phoresis** (literally transmission), as the name suggests, is an association based on transportation.

Dung pats in pasture constitute a rich habitat for many species of animal but they are temporary and liable to dry out or disintegrate and be washed away, depending on the prevailing weather. Their temporary nature is no problem to many of their inhabitants that are actively motile and can migrate to a fresh pat. For less active species the loss of habitat may be more serious and they may have to rely on haphazard contamination of the surrounding

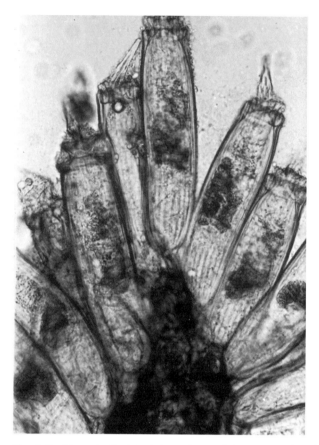

Fig. 1.3 Phoresis. Resistant dauer larva of the nematode *Pelodera coarctata*. The nematodes live on bacteria in faecal pats. As the habitat deteriorates they enter this resistant dormant state and attach to the cuticle of dung beetles for transport to fresh faeces.

pasture to effect transmission. The free-living nematode *Pelodera coarctata* solves the problem of transfer by 'hitching a lift' on another inhabitant of the dung pat, the dung beetle, *Aphodius fimentarius*. Normally the nematodes live and breed in the dung, feeding on the plentiful supply of bacteria there. As the dung begins to dry out they produce special desiccation resistant stages known as the **dauer larvae** (Fig. 1.3). These attach to the body of the beetle and are transported to a new site. As both nematode and beetle have a common aim in reaching fresh dung the partnership works well; there is no physiological connection between them and the association is only temporary.

Lice are wingless ectoparasitic insects that feed on blood, tissue debris and feathers of their hosts. Transmission of lice from host to host depends on close contact as the entire life-cycle is spent on the host. Hippoboscid flies, **keds**, are ectoparasites on a range of birds and mammals, feeding on their blood. Many species of ked lose their wings once they have located a host but others remain winged. They generally breed in the host's nest so that each generation of hosts is infested early in life. The most disastrous event in the life of a louse is the death of its host. It is unable to fly and has no obvious means of locating a new host. Here the ked may come to the rescue. *Sturnidoecus strurni* is a louse specific to starlings, and on a live starling it ignores keds (*Ornithomya fringillina*). If the bird dies either the keds become more attractive or less repellent than the cooling flesh of the bird and the lice attach to the keds which carry them to another host. The keds are not especially host-specific but the lice are so there is always a risk that the ked will attach to an unsuitable host. However, starlings are gregarious birds and the chance of the louse arriving on another starling is quite high. As both lice and keds are regarded as parasites when on the host it is apparent that how they are defined depends on the relationship that they are in at any given time.

Synoecious commensalism **Synoecious** (*syn* – together, *oikos* – house) **commensalism**, sometimes called **inquilism** (from the Latin *incolinus* – who lives within) is an association in which one animal, usually smaller and more motile, lives within the shell or body of another. There are numerous examples of animals that are regularly found together. Nereid worms may occupy the shells of hermit crabs and remove food from the mouth-parts of their hosts. The pea crab, *Pinnotheres pisum* (Fig. 1.4), exclusively lives within the mantle cavity of bivalve molluscs, especially mussels, *Mytilus edulis*, and cannot live independently. There is little evidence that these 'lodgers' cause any harm to their 'landlords' so the benefit is unilateral.

The fish *Fierasfer* lives within the body cavity of sea cucumbers; it enters the cloaca, despite some resistance, to find shelter. There is some debate about the effect of the fish on the holothurian. Some authors consider that the fish feeds on crustaceans and causes no damage to its host while other authors consider that it feeds partly at the host's expense and, if this is true, it is a very small conceptual step from this association to parasitism.

Parasitism (literally *para* – beside, *sitos* – food)

Over the years there have been innumerable different definitions of parasites, parasitism and the parasitic relationship. At least one parasitologist has defined parasites as the organisms studied by parasitologists! This is neither

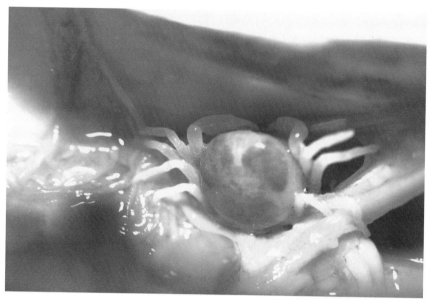

Fig. 1.4 Synoecious commensalism. A female pea crab, *Pinnotheres pisum*, in the mantle cavity of its mussel host. The crab does not damage the mussel and uses its shell purely for protection.

informative nor helpful but it does illustrate the point that there is no universally accepted definition. Part of the problem is that parasitology itself encompasses a wide range of scientific disciplines, from the world-wide view of disease and disease transmission studied by epidemiologists to the submicroscopic DNA analysis within the cells of individual parasites, and includes immunology, pathology, cytology, biochemistry, physiology, behaviour, genetics, ecology, evolutionary theory, mathematical analysis and human and animal medicine, amongst other topics. There is at least one school of thought that suggests that all biology could be taught through the medium of parasitology!

As most of the rest of this book is about parasites we do need a working definition, and for our purposes I suggest that we consider parasites as being animals that live for an appreciable proportion of their lives in (**endoparasites**) or on (ectoparasites) another organism, their host, are dependent on that host and benefit from the association at the host's expense. This definition incorporates the idea that a parasite damages its host. This is not always easy to demonstrate as an individual parasite may cause no recognisable damage: disease is frequently a population phenomenon and even when damage does occur it is often as a result of something else in the host/parasite relationship going wrong (see Chapter 6). The definition also includes the idea that parasites are dependent on their hosts, and this obligatory nature of the relationship distinguishes parasites from some of the other associations we have already considered. For the moment this definition will differentiate the parasites from the other groups of symbionts and its deficiencies will become apparent later.

Mutualism

The term mutualism is often used rather loosely to cover any association where both partners benefit. The hermit crab/sea anemone association which, while mutually beneficial, is not obligate; and the synoecious commensal mussel/pea crab relationship, where the benefit is unilateral, are regarded as mutualism by some authors.

True mutualism, where there is an obligate, mutually beneficial association between two different species is relatively rare amongst animals. The classic examples of mutualism in this restricted sense are found amongst the protozoan inhabitants of wood-eating insects and ruminants.

There are some 2000 species of termites throughout the tropics. These are colonial, social insects which are best known for the elaborately sculpted mounds that many build. Most species feed on wood, leaves and other plant material although some cultivate fungi for food. The colony consists of a grossly enlarged egg-laying queen who is attended by a single male, the king, and large numbers of workers, all protected by specialised soldiers. It is the worker caste that is responsible for collecting the food and passing it to the other members of the colony. The well-being of the whole colony thus depends on the nutritional physiology of the workers.

The hind gut of termites has an enlarged sac, the paunch, which contains vast numbers of flagellate protozoans and bacteria. These organisms may account for up to one-third of the total weight of the insect. The paunch is anaerobic and the protozoans **phagocytose** fragments of plant material and degrade them to hydrogen, carbon dioxide and fatty acids, especially acetic acid. The acids are absorbed through the gut wall and metabolised aerobically.

Selective elimination of flagellates has demonstrated that not all are equally important. However, if they are all removed and the bacteria left unaffected the termite will continue to feed on wood but cannot digest it and eventually dies of starvation. The exact role of the bacteria remains uncertain: they may provide food for the protozoans, they may help to maintain anaerobic conditions in the paunch, they may provide additional enzymes to partially digest

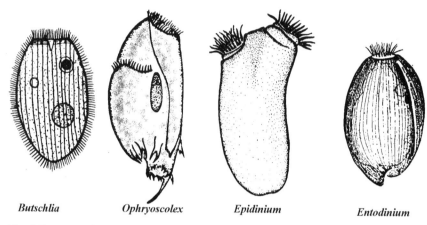

Butschlia   *Ophryoscolex*   *Epidinium*   *Entodinium*

Fig. 1.5 Mutualism. A selection of ciliates from the rumen. The rumen contains enormous numbers of ciliates and bacteria that break down cellulose in the feed. The ciliates belong mainly to two subphyla, the Holotrichia with even distribution of cilia and the Peritrichia with compound cilia.

the food or they may fix atmospheric nitrogen and recycle the nitrogenous waste products of their hosts.

As the paunch constitutes part of the hind gut its lining is lost each time the termite moults and exposure to atmospheric oxygen kills its contents. Termites practise what to us may seem a rather unsavoury habit called **proctodeal feeding**. They expel a drop of fluid from the hind gut which is ingested by another insect. This drop of fluid contains some of the contents of the paunch, which is passed from one instar to the next as well as from generation to generation as the gut of newly hatched larvae needs to be infested before they start to feed.

Not only are the enormous termite mounds that are such a feature of the landscape in parts of the tropics dependent on microscopic protozoan mutualists but so also is a considerable amount of the world's agriculture.

The stomach of ruminants, including cattle and sheep, consists of four chambers, the rumen, the reticulum, the omasum and the true stomach, the abomasum. Food comes into the first of these after mechanical breakdown by the chewing and grinding action of the teeth. Following a period of fermentation in the rumen the food is regurgitated and chewed again to break it down into even smaller particles, with consequently greater surface area – the process of 'chewing the cud'. The rumen is an anaerobic fermentation chamber that contains the finely ground plant material together with 10^5 ciliated protozoans (Fig. 1.5) and 10^9–10^{10} bacteria per ml. The end products of fermentation of plant material are largely fatty acids but the pH of the abomasum remains approximately neutral because of the buffering activity of the bicarbonates and phosphates in the vast amounts of saliva that are swallowed with the food. This contributes to the liquid phase of the rumen contents. The gaseous phase consists largely of carbon dioxide and methane: any oxygen that may be taken in is rapidly dissolved and metabolised by facultative aerobic bacteria. Excess gas is expelled by belching.

The ciliates phagocytose plant fragments and break them down into their constituent sugars or ferment them to fatty acids which are absorbed by the animal. Different measurements suggest that in the normal rumen some 30–60% of fermentation is due to the ciliates. However, if the ciliates are removed by chemical treatment the animals are able to grow and perform in a manner comparable to untreated controls. This suggests that the protozoans are important but not necessarily essential to the animal. Each day about a third of the rumen inhabitants are passed from the rumen into the reticulum and thence to the abomasum where they are digested, so they may contribute to the nutrition of the host a second time.

The gut of new-born ruminants is essentially sterile but as soon as they begin to take solid food the gut rapidly becomes colonised by the necessary micro-organisms. Most of these come from saliva when the young animal is licked by its mother but some of the smaller organisms may come from aerial droplet transmission.

At this stage it is perhaps easy to sympathise with Disraeli (1804–1881) when he stated '*I hate definitions*'.

2

Parasites

Parasites generally receive a 'bad press'. Students, the unemployed and many other misunderstood minorities are referred to as 'parasites' as a term of abuse. However, parasitism is a widespread and respectable way of life and far from being degenerate most parasites are highly and beautifully adapted to link their life-cycle to that of their host. The fact that at least two distinct types of organism, the parasite and the host, are involved adds greatly to the complexity of the relationship and also to the fascination of unravelling and understanding it. In addition, some parasites are of economic significance from the diseases they cause to humans, their animals and their crops and this gives their study additional importance and value.

It seems rather unjust that predators that aim to kill their prey should be regarded as commendable and somehow noble animals while parasites, which often cause little or no detectable harm to their hosts, should be abhorred. It is not true, as we shall see later, that a parasitic infection inevitably results in disease symptoms and, as some of the most beautifully engineered and adapted of all animals, parasites deserve our admiration rather than our revulsion.

Parasitism is not an unusual way of life and has evolved many times. With the possible exception of the echinoderms there are examples of parasites in all the major and most of the minor phyla. The lack of a parasitic echinoderm may be more a reflection of a lack of knowledge of the phylum rather than a true statement that no such example exists. As far as numbers of organisms is concerned the number of parasites greatly exceeds that of free-living animals. Every free-living species that has been examined in any detail has a number of parasites associated with it. Some species may have a potentially enormous parasite fauna, e.g. 151 different **helminth** (parasitic worm) species have been recorded from herring gulls (*Larus argentatus*) and 342 from man, although it must be admitted that some of the latter are of somewhat dubious validity.

Every parasite must have at least one host in its life-cycle and many species have more than one. The host in which the parasite spends most of its time and usually in which sexual reproduction occurs is known as the **definitive host**. If there is only a single host in the life-cycle then transmission is said to be direct and the parasite to have a **direct life-cycle**. Many parasites have more complicated life-cycles with additional hosts in which development of the parasite proceeds; these other hosts are known as **intermediate hosts** and, if they are especially motile and actively carry the parasite to another host, as **vectors**. Thus a vector is an intermediate host with special behavioural attributes: not all intermediate hosts are necessarily vectors. Life-cycles with more than one host are known as **indirect**. There are some parasites that make use of additional hosts to overcome adverse environmental conditions and to extend their life-span. No development of the parasite occurs in these hosts so the developmental stage of the parasite that enters is the same as that which leaves. These hosts are known as **paratenic** or **transport hosts**.

While most phyla contain a few species that have adopted a parasitic way of life a few phyla contain a large proportion of the economically important parasites. Traditionally these have been studied by three different groups of investigators and have their own specialised literature. However, many of the principles of parasitism are common to all groups and it is desirable to gain an appreciation of the parasites as a whole before considering specialisation.

Protozoa

Firstly there are the single-celled animals, the Protozoa, the province of the protozoologist. There has been considerable debate about the position of the Protozoa in the phylogeny of both the animal and plant kingdoms. Living organisms are divided into groups, or taxa, on the basis of shared characteristics. There are many possible characteristics that could be used as the basis for classification systems, for example, one could select colour or whether the organisms are good to eat as criteria. However, the aim of systematists is that their classification should have a sound scientific basis and should reflect evolutionary changes. Where there is a fossil record evolutionary trends may be recognisable, but for many of the smaller soft-bodied organisms no such record exists and classification depends on the detailed examination of living representatives. As knowledge increases and new techniques for examination become available ideas of the most significant characteristics change and new relationships are revealed. The basic building block of any of the taxonomic systems is the species as this is the only unit that has a genuine biological definition: members of a species are capable of interbreeding and producing viable, fertile offspring. The remainder of the taxonomic edifice is man-made and, while it is hoped that it is constructed on sound evolutionary foundations, is subject to continual alteration. There has been considerable debate over the last 20–30 years about the most appropriate classification system and this has perhaps had a greater effect on the classification of the Protozoa than on any other of the animal groups. Firstly the electron microscope and more recently genetic and molecular biological techniques have changed how protozoologists have looked at their organisms. There have been a number of major revisions of the classification of the Protozoa and the whole question is under constant review.

Most authorities now accept at least a third kingdom, the **Protista**, that encompasses both the animal and plant-like single-celled organisms. Some taxonomists advocate up to 19 kingdoms! As we shall be considering animal parasites we will concentrate exclusively on the Protozoa. The four traditional divisions of the Protozoa were based largely on major morphological characteristics, but there seems little doubt that each of these contained a number of phyla. It has been proposed that the flagellated protozoa alone may come from up to 14 different phyla. This is a complex area and for our purposes a much simplified version will be adopted dividing the Protozoa into four Groups of phyla based largely on their means of locomotion. This is in line with current systematic thinking and has much in common with traditional classification.

The first Group contains the amoeboid protozoa with a single nucleus and locomotion by **pseudopodia**. They are widely distributed in terrestrial, freshwater and marine environments; the chalk downlands of southern England are largely composed of the calcareous tests of countless millions of amoeboid foraminiferans of the Cretaceous period around 100 million years ago.

It is perhaps not surprising that animals that can adapt to such a wide range of different free-living environments should also have parasitic relatives and, while there is not a great number of parasitic amoebae, some, e.g. *Entamoeba histolytica* (Fig. 2.1), the cause of amoebic dysentery in man, may be important. Parasitic amoebae live in the gut of their hosts and reproduce by binary fission. All of the parasitic species additionally form cysts to provide protection from the rigours of the environment during passage from one host to another. Transmission is thus by contamination of food, water or fingers with faecal material from an infected person. This type of direct life-cycle is thus often known as **contaminative** or **faeco-oral transmission**.

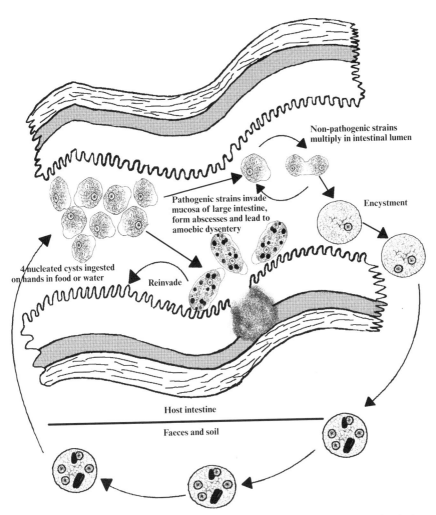

Non-pathogenic strains
multiply in intestinal lumen

Pathogenic strains invade
mucosa of large intestine,
form abscesses and lead to
amoebic dysentery

Encystment

4 nucleated cysts ingested
on hands in food or water

Reinvade

Host intestine

Faeces and soil

Fig. 2.1 The life-cycle of *Entamoeba histolytica*. Infection occurs contaminatively from the ingestion of cysts which excyst in the intestine. Trophozoites reproduce by binary fission in the lumen. Some strains are pathogenic and invade the mucosa causing abscesses when they may enter the blood stream and may be carried to other sites, especially the liver, to set up additional abscesses.

Like the amoeboid protozoa many of the second Group of flagellated protozoa are free-living, but it also includes some important and widely distributed parasites. The major distinguishing feature of the flagellates is the possession of one or more **flagella** (singular flagellum) (Fig. 2.2). These lash-like structures are used for locomotion. Reproduction is normally by binary fission but there is evidence of genetic exchange and hence sexual reproduction

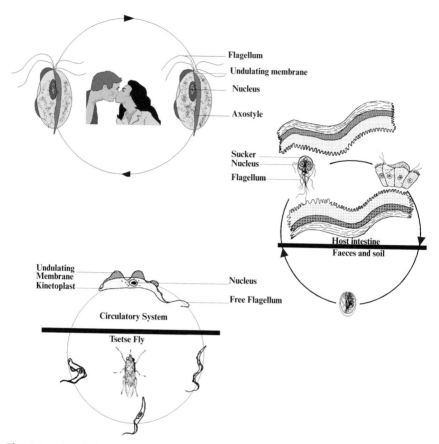

Fig. 2.2 Flagellate protozoan parasites. *Trichomonas* (top) does not form cysts and relies on intimate contact for transmission. *Giardia* (centre) lives attached to the surface of duodenal cells and is transmitted contaminatively in cysts. *Trypanosoma* (bottom) lives in the blood and is transmitted by the bite of the tsetse fly.

amongst some. Parasitic flagellate life-cycles and sites in the host are more varied than the amoebae and both gut and tissue sites may be parasitised by different species. *Trichomonas* spp. do not form cysts and are transmitted directly from host to host by intimate contact, genital in the case of *T. vaginalis* and osculatory in the case of *T. tenax*. *Giardia* produces cysts and transmission is contaminative while the trypanosomes use an insect vector to transmit them from the closed circulatory system of one host to the next.

There are some protozoans that share the characteristics of both the amoeboid and the flagellate protozoa. For part of their life-cycle they are amoeboid and for part flagellated. The phylogenetic relationships of this Group have

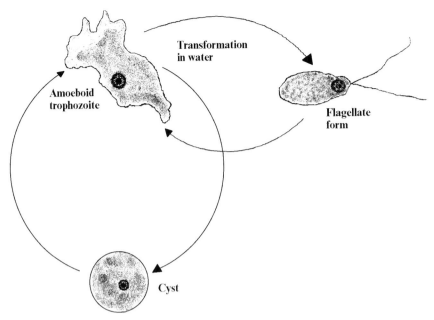

Fig. 2.3 *Naegleria fowleri*, a protozoan that has both amoeboid and flagellate forms. It normally lives in the soil and infects man when forced into the nasal passages when swimming.

always been problematical and some classifications of the Protozoa have combined the amoeboid and flagellated protozoans into a single group, the Sarcomastigophora, to accommodate them. More recent classifications have again divided the two Groups but there is still dispute as to which Group these amoebo-flagellates share most allegiance with. One such example is *Naegleria fowleri* (Fig. 2.3). This species normally lives in the soil as an amoeboid **tropho-zoite** (feeding stage) dividing by binary fission and encysting when environmental conditions deteriorate. If the soil is diluted with water they transform into flagellated forms. If these are forced deep into the nasal passages of man while swimming in infected water they migrate along the olfactory nerve into the cranium and may result in rapid and fatal damage to the brain. Isolated cases have been reported from many parts of the world and the species are widespread. The risk of being infected is however very small.

The third Group contains the ciliated protozoa. The ciliates are typified by either simple **cilia** or compound ciliary organelles and their associated infra-ciliature. They normally have two types of nuclei and are very largely free-living. Mutualistic rumen ciliates have already been discussed and the only ciliate regularly reported as a parasite of mammals is *Balantidium coli* (Fig. 2.4).

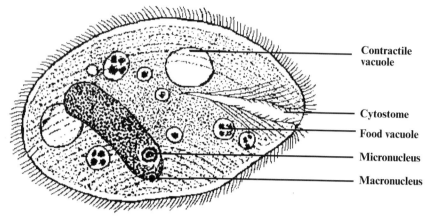

Fig. 2.4 *Balantidium coli*, a ciliate parasite with a direct contaminative life-cycle from the intestine of man and pigs.

This is a large ciliate (30–130 μm long × 20–120 μm broad) and is a common parasite of the caecum and colon of pigs and occasionally man. It normally occupies the gut lumen but in man may invade the mucosa and cause dysentery. The parasite reproduces both by binary fission and conjugation and forms cysts that are passed in the faeces. Infection of man normally results from contamination of food or water with pig faeces although man to man contaminative infection has been reported.

The last Group is the **Sporozoa** which differ from the other protozoans in lacking obvious locomotory structures and in having, at some stage in their life-cycle, a number of organelles at one end of the cell known as the apical complex. An alternative name for this Group is the **Apicomplexa.** The apical complex is used to ease the passage of the parasite into host cells but, as it is only visible at the electron microscope level of magnification, it is of little value in recognition or identification. It is, however, an integrating feature that has been helpful in resolving taxonomic questions. From a parasitologist's point of view this Group is arguably the most important of the protozoan Groups as all its members are parasites. Many sporozoans are intracellular parasites and have complex life-cycles with alternating asexual and sexual generations. The result of sexual reproduction is often a resistant spore which accounts for the name of the Group. The Group is now divided into a number of classes, including the Coccidia, e.g. *Eimeria*, which form cysts and have a direct, contaminative life-cycle, and the Haemosporidia, e.g. *Plasmodium*, which do not and have an indirect life-cycle with an arthropod intermediate host. The essential features of the life-cycle of both classes are similar, with differences related to transmission.

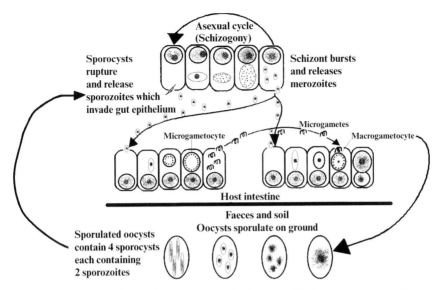

Fig. 2.5 Life-cycle of *Eimeria*, a contaminatively transmitted sporozoan parasite. Following ingestion of the sporulated oocysts the parasite undergoes a number of asexual generations before producing sexual stages and cysts.

Eimeria (Fig. 2.5) and related genera are often known as **coccidian** parasites. They are intracellular parasites especially of the intestinal tissues of vertebrates and are very widely distributed. Most vertebrates may be infected by at least one coccidian. The host is infected following ingestion of oocysts which excyst during passage through the gut releasing sporozoites which invade intestinal epithelial cells. Here they undergo an asexual reproductive process known as **schizogony**. The nucleus divides several times resulting in an enlarged multinucleate host cell. The nuclei move to the edge of the parent cell, the **schizont**, and daughter individuals, **merozoites**, bud off with one to each nucleus. Eventually the cell bursts, destroying the schizont and releasing the merozoites. These invade further cells and may undergo further schizogony. Instead of producing schizonts some of the merozoites form either multinucleate **microgametocytes** or uninucleate **macrogametocytes**. When mature the microgametocytes burst releasing into the gut lumen flagellated **microgametes**. These seek out the **macrogametes** which have developed in the macrogametocytes and fertilise them. The fertilised zygote secretes a protective wall around itself and becomes an **oocyst**. The process of sexual reproduction is known as **gametogony**. An oocyst has to undergo a period of maturation and development known as **sporogony** before it will be infective to another host. In the case of *Eimeria* the zygote divides three

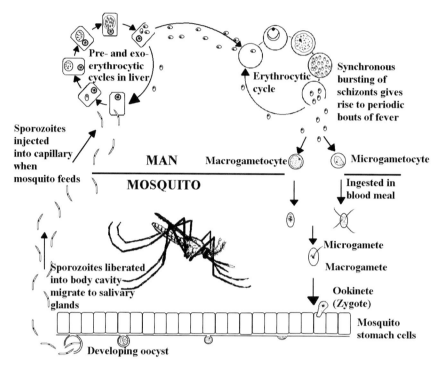

Fig. 2.6 Life-cycle of the malaria parasite *Plasmodium*. The life-cycle is essentially that of *Eimeria* with the asexual stages in man and maturation of the sexual stages in a mosquito vector.

times resulting in eight **sporozoites** which, in a fully sporulated oocyst, are arranged in four pairs within protective **sporocysts**. Related species have different arrangements, *Isospora* species, for example, have two sporocysts each containing four sporozoites in their oocysts.

The life-cycle of the malaria parasite, *Plasmodium* (Fig. 2.6) has many features in common with *Eimeria*. The infective stage is a sporozoite but this is injected into the bloodstream when the host is bitten by an infected mosquito. The sporozoites initially enter liver cells and undergo **pre-erythrocytic schizogony**. The merozoites released either reinvade liver cells where they may lie dormant and may be responsible for the relapses that are typical of malaria, or more commonly pass into the blood stream and enter red blood cells. In the blood they undergo a series of **erythrocytic schizogonies** until a large number of red cells become infected. These asexual reproductive cycles become synchronous and the periodic bouts of fever characteristic of malaria result from the almost simultaneous bursting of many schizonts and the release of the parasites' metabolic products into the blood stream. The different frequencies of fever in different malaria species reflect the developmental times of the schizonts. The development of sexual stages occurs when merozoites develop into macrogametocytes and microgametocytes rather than schizonts. The gametocytes circulate in the blood and do not develop further unless they are taken up by a female *Anopheles* mosquito when she takes a blood meal. Within the mid-gut or stomach of the mosquito the nucleus of the microgametocyte divides three times, the nuclei move to the periphery of the cell and eight flagella emerge from the surface. One nucleus moves to each flagellum to form a microgamete that separates from the gametocyte and passes into the gut lumen. The microgametes fuse with the macrogametocytes to form a motile zygote called an **ookinete**. Meiosis occurs at the first nuclear division of the zygote so that for most of the life-cycle the parasites are **haploid** and only for a transitory period **diploid**. The ookinete penetrates the epithelium of the mosquito gut and settles under the basement membrane on the body cavity side. Here it grows as an oocyst. The nucleus divides repeatedly followed by the cytoplasm and eventually the oocyst bursts to release hundreds of sporozoites into the body cavity. These migrate anteriorly and enter the salivary glands of the mosquito where they wait until the mosquito bites another host to be injected into the blood stream.

Helminths

The worm-like parasites, the helminths, are contained in three phyla, the **Platyhelminthes** or flatworms, the **Nematoda** or roundworms and the **Acanthocephala** or spiny-headed worms. These are the province of the helminthologist and again there is a specialised vocabulary and literature associated with helminthology. While some representatives of the Platyhelminthes and Nematoda are free-living, the Acanthocephala are exclusively parasitic. The helminths are multicellular and, while some of the smaller parasites in each phylum may be little larger than some large protozoans, they are generally larger and some may grow to relatively massive sizes, several metres long in the case of cestodes and nematodes.

Platyhelminthes

The phylum Platyhelminthes contains four main classes and a number of minor ones. Of the main classes the **Monogenea**, **Digenea** and **Cestoda** are exclusively parasitic while the Turbellaria are mainly free-living with a few ectoparasitic representatives. The different classes show marked differences but share features that allocate them to a single phylum. They are dorso-ventrally flattened acoelomate metazoans, that is they are flat multicellular worms without a body cavity, the internal organs surrounded by a spongy parenchyma which packs the space within the body wall. The body wall varies between the different classes (see Fig. 5.12) but in each case consists of a syncytial tegument which may be absorptive, secretory and protective. The alimentary canal, if present, is simple and blind, ending without an anus, and the excretory system consists of flame cells and collecting tubules which open to the outside. The great majority have a complex hermaphrodite reproductive system. At one stage the Monogenea and Digenea together with one of the minor platyhelminth classes the **Aspidogastrea** were considered to be closely related and were put together in a single group, the **Trematoda**. More recent work has suggested that the links between the classes are only superficial and the Trematoda is no longer seen as a viable taxonomic grouping. However, the term trematode, usually applied to a digenean, may still be encountered in the literature.

Monogenea

The Monogenea (Fig. 2.7) are flattened leaf-like organisms that are typically ectoparasites of fish, living either on the skin or the gills. A few species have invaded the cloaca and bladder of amphibians and reptiles and one species has been recorded from the eye of the hippopotamus. They vary in size from about 0.5 mm to 30 mm and differ from the other platyhelminths in having a posterior attachment organ, the haptor or **opisthaptor**, which may be armed with hooks and suckers and show considerable adaptation for attachment to specific sites on the host. The life-cycle is simple and direct and depends on water for transmission. Cross-fertilisation between two hermaphrodite individuals appears to be common but is unlikely to be invariable. Most species lay **operculate** eggs which hatch in water to give free-swimming **oncomiracidia**. These locate a new host, attach and mature to adults. While this life-cycle is superficially simple the monogeneans exhibit a range of adaptations to encourage contact between oncomiracidia and appropriate hosts (see Chapter 3). Few monogeneans cause serious damage to their hosts: those that live on the skin feed mainly on mucus and epithelial cells, although some may also feed on blood. The gill and bladder species also feed on blood

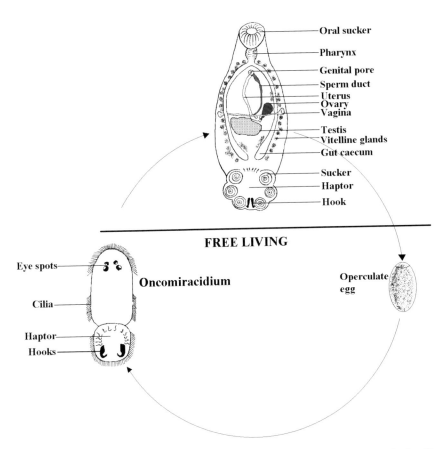

Fig. 2.7 General monogenean life-cycle. The hermaphrodite adults are typically gill or skin parasites of fish. Operculate eggs hatch to release oncomiracidia that seek out and attach to a new host.

but the population level is rarely great enough for this to affect the health of the host. In a few species the egg hatches in the uterus of the female and an immature adult, itself containing another developing embryo, is released. These species, e.g. *Gyrodactylus*, can result in a very rapid population build-up and can cause considerable damage to young fish, especially under the crowded conditions of commercial fish hatcheries. *Gyrodactylus salaris* has been a cause of considerable concern to salmon farmers in Norway in recent years where it has spread rapidly with the growth of fish farming. A major eradication campaign which depends on killing all the fish in infected rivers has been introduced and it is hoped that this will be complete by the end of the century.

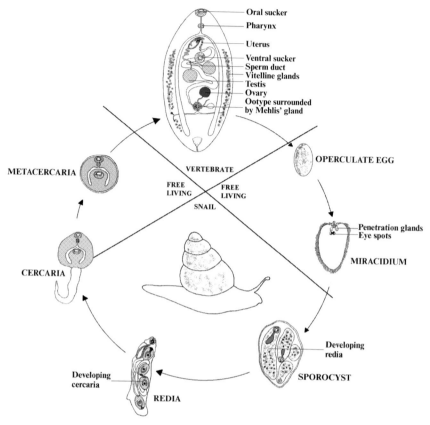

Fig. 2.8   Digenean life-cycle. The adults are endoparasites of vertebrates. The eggs hatch to release miracidia which invade a mollusc. There are generally two stages of asexual reproduction resulting in cercariae that leave the snail to encyst as metacercariae to be ingested by the definitive host.

Digenea

As adults the digeneans, or **flukes** (Fig. 2.8), are almost invariably endo-parasites of vertebrates (one or two ectoparasitic species are known) and vary in size from about 0.5 mm to 100 mm. All classes of vertebrates may be infected and different digenean species may inhabit a wide range of organs including the gut, liver, lungs and circulatory system. Unlike the monogeneans the digeneans have no opisthaptor and attachment to the host is by **suckers**. Most species have two suckers: one, the oral sucker, is anterior and surrounds the mouth, the other, the ventral sucker, is more variable in position and, in different species may occupy almost any position on the ventral surface. A few species lack a ventral sucker and some have a third, genital sucker, surrounding the genital opening. The presence, size and position of the suckers are often useful distinguishing features between species.

The reproductive system is generally complicated and hermaphrodite although cross-fertilisation commonly occurs. One important group of digeneans, the schistosomes, is dioecious but the males and females live permanently paired and the females only mature to produce eggs when in association with a male.

The life-cycle of digeneans is indirect. The first intermediate host is almost invariably a mollusc and commonly a gastropod. Digeneans show a much greater specificity for their intermediate hosts than they do for their definitive hosts. It has been argued that this reflects the evolutionary development of the digeneans from free-living predatory turbellarians *via* ancestral mollusc parasites to the present day digeneans with sexual reproduction in vertebrates and asexual multiplication in molluscs.

To achieve the transfer from host to host most digenean life-cycles include two free-living stages. Eggs, which usually have an operculum, are the product of sexual reproduction in the vertebrate host. They are released, generally in the faeces, and hatch in the environment to release free-swimming ciliated **miracidia**. These have a limited life-span, measured in hours rather than days, during which they have to locate and invade a suitable mollusc intermediate host. Once inside the mollusc they shed the ciliated coat and transform into a **sporocyst**. This consists of a rather structureless sac of germ cells derived from the miracidium which develop into the next stage of the life-cycle, the **redia**. The rediae are more structurally defined and have a mouth and short blind-ending gut. They are capable of moving through the tissues of their hosts. Within the rediae the last stage of the life-cycle, the **cercariae**, develop. These are small motile stages which leave the mollusc, swim with the aid of their muscular tails and encyst as **metacercariae** on vegetation, or on or in a second intermediate host. It is these metacercariae which are infective when ingested by the vertebrate host. The asexual reproductive amplification in the mollusc is enormous and a single miracidium may give rise to 500 or more cercariae. The digenean life-cycle is highly adaptable and different species illustrate many different modifications on the basic theme (see Chapter 3).

Cestoda

For many people the tapeworms represent the archetypal parasite; long and sluggish, they rely on their host for warmth and protection and lie bathed in a soup of partially digested nutrients which they absorb directly without having to do any apparent work for their living. In fact the cestodes are the most specialised of the parasitic helminths and are superbly adapted to their habitat and way of life.

There is evidence that both monogeneans and digeneans may be able to absorb some nutrients across their body wall to supplement those obtained from the gut, but for the cestodes the tegument is the sole absorptive surface and is thus more complex than that of the other platyhelminths. It has to serve protective, secretory and absorptive functions (see Chapter 5).

Because the cestodes lack any sort of gut of their own they are dependent on their hosts for the initial breakdown of large nutrient molecules and this restricts the sites in the host that they can inhabit. As adults, tapeworms are typically parasites of the intestinal tract of vertebrates with a few species inhabiting ducts, e.g. the bile duct and pancreatic duct associated with the intestine. All classes of vertebrate may be infected but only two species are known that mature in invertebrates.

A tapeworm (Fig. 2.9) consists of an anterior attachment organ, the **scolex** or holdfast, and a more or less elongate flattened ribbon, the **strobila**, composed of individual segments or **proglottides**, each of which contains a complete hermaphrodite reproductive system. The Cestoda are divided into a number of orders which are distinguishable by the structure of the scolex. Most of the orders contain mainly fish parasites but two, the Pseudophyllidea and the Cyclophyllidea, contain parasites of warm-blooded vertebrates. Pseudophyllidean cestodes possess a scolex with two sucking grooves or **bothria**, while the scolex of cyclophyllidean cestodes has four suckers arranged around it and often a **rostellum of hooks**. The proglottides are not independent individuals but are linked through a common nervous system and excretory system to one another. The proglottides form just behind the scolex in an unsegmented area known as the **neck**. As the proglottides mature others form in front of them and they gradually move down the length of the strobila. In each proglottid the male system usually develops first and as they get older and pass further down the strobila the female system matures. Fertilisation occurs between worms, between proglottides of the same worm and possibly self-fertilisation of the same proglottid to produce fertilised eggs. Depending on the type of cestode the eggs may be laid into the gut of the host to be voided in the faeces or they may be retained in the uterus which enlarges to accommodate them. When fully gravid the whole segment is then lost from the posterior of the worm and leaves the host as a package of eggs. Although not as active as some other helminths the cestodes are capable of co-ordinated movement using muscles in the body wall and may undergo migrations in the intestines of their hosts (see Fig. 5.2).

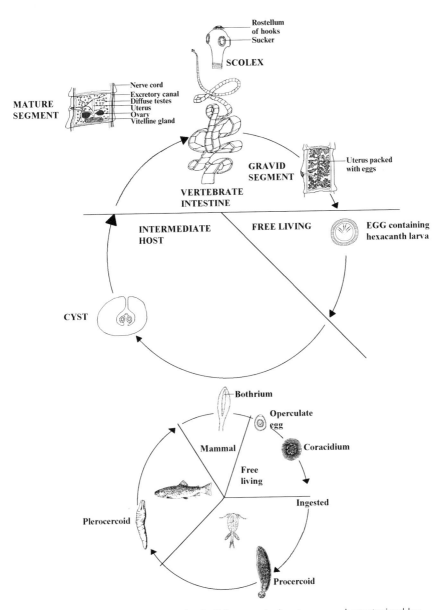

Fig. 2.9 Cestode life-cycles. Cyclophyllidean cestodes, top, are characterised by having a scolex with four suckers and a strobila of proglottides of increasing maturity which, when gravid, are shed in the faeces. The single intermediate host is infected by eating the eggs and the cystic stage forms in the tissues. Pseudophyllidean cestodes, below, have a scolex with two sucking grooves, bothria, and many of the proglottides are at a similar state of development, releasing eggs rather than gravid segments. The life-cycle involves two intermediate hosts, the first an invertebrate and the second a vertebrate.

Cestodes have indirect life-cycles (Fig. 2.9) and the fate of the eggs depends on the order of tapeworm concerned. Pseudophyllidean cestode eggs hatch to release a motile ciliated **coracidium** larva which swims freely in water until it dies or is ingested by a suitable intermediate host, typically an arthropod, e.g. *Cyclops*. Within the arthropod it penetrates the gut wall and passes into the haemocoel where it elongates to form a **procercoid**. The procercoid only develops further if its host is eaten by a second intermediate host, commonly a fish. Within the body cavity of the fish the procercoid elongates and develops into a **plerocercoid** which is the infective stage for the definitive host. Thus the pseudophyllidean life-cycle, like that of monogeneans and many digeneans with free-swimming larval stages, is dependent on water for transmission.

The cyclophyllidean cestodes have eliminated the need for water. The eggs are infective as they leave the worm and are ingested by the intermediate host, which for different tapeworms may be either an invertebrate or a vertebrate. The eggs hatch in the intermediate host's gut to release motile **oncospheres**: these take the form of **hexacanth** (six-hooked) larvae which penetrate the gut wall and develop into a cystic stage in the tissues. The type and size of the cyst varies from species to species. The definitive host is then infected when it eats the cyst, usually with the intermediate host. Thus the typical cyclophyllidean cestode life-cycle has the adult worm in a carnivore and the intermediate stages in a herbivore. Omnivores get the worst of both worlds being able to act as both definitive and intermediate host, sometimes, e.g. *Taenia solium* in man, for the same worm. This does not mean that herbivores are exempt from tapeworms. Many cyclophyllideans use invertebrates as their intermediate hosts and these may be ingested by herbivores while grazing.

Nematoda

There is a malicious rumour perpetuated by those who know no better that having seen one nematode you have seen them all. Like all good rumours there is an element of truth in it as nematodes do show marked conservatism in structure and, with very few exceptions from the smallest, about 250 μm long, to the largest, *Placentonema gigantissima* which is about 9 m long, all have a very similar body plan. While they exhibit structural constancy the nematodes show enormous biochemical plasticity and are amongst the most ubiquitous of all animals. Nematodes have been recorded from Antarctic ice and hot springs, there are soil, freshwater and marine free-living nematodes and parasites of animals and plants. There is one species that lives in vinegar and another that inhabits beer mats. Indeed it has been suggested that if all the matter were removed from the earth except the nematodes we would be left with a filmy outline of everything. It has to be admitted that this idea was proposed before the advent of chipboard and modern plastics but it does offer a graphic if slightly fanciful illustration of nematode ubiquity.

It is often claimed that we live in the age of the insects, as more insect species have been recognised than any other type but, as all insects that have been examined in detail have at least one nematode parasite, a good case can be made for this being the age of the nematode.

Most nematodes are small and inconspicuous. The estimated number in an acre of agricultural land varies from 3–9,000,000,000 and in an acre of marine beach sand about half this number. However, few people ever see them or realise they are there. Some of the large human parasitic nematodes were known to the Egyptians about 1500 BC and the fiery serpent that was sent to plague the Children of Israel (about 1250 BC) is thought to have been the Guinea worm (*Dracunculus medinensis*), but most knowledge of nematodes is recent. The importance of plant parasitic nematodes in causing crop losses only became apparent after World War II when effective insecticides became available to reduce the more conspicuous insect pests. Production did not increase as much as had been anticipated and the nematodes were recognised as serious pathogens. The damage they do their plant hosts may be direct or they may introduce viruses or cause lesions which are subject to secondary bacterial or fungal invasion. The estimated annual financial loss due to nematodes for major world crops in the mid-1980s was over £30,000,000,000.

The basic nematode body plan consists of two concentric tubes (Fig. 2.10). The outer consists of the cuticle which is secreted by the underlying hypodermis, and the longitudinal somatic musculature. The inner tube consists of the gut. The space between the two is known as the **pseudocoelom** (it is not a true coelom as its origin is the primary cavity of the embryo while a coelom is derived from a secondary cavity), and is fluid filled. The pressure developed in this fluid acts as a **hydrostatic skeleton** and helps to counter the contraction of the longitudinal muscles to allow sinusoidal locomotion without the need for antagonistic circular muscles.

Unlike the platyhelminths the nematodes have a through gut with both mouth and anus. The mouth is anterior and usually terminal. The shape of the buccal capsule and other oral structures are features that help to distinguish one nematode from another. Plant parasitic nematodes usually have a cuticular **stylet** with which they penetrate the cell walls of their host, while animal parasites may have a number of cuticular adaptations including teeth and cutting plates to breach their host's tissues.

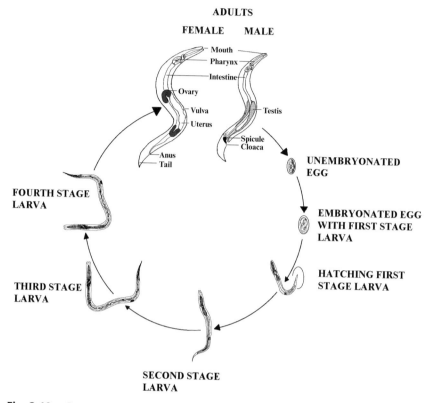

Fig. 2.10 General nematode structure and life-cycle. The dioecious adults produce eggs that hatch to release larvae. There are four larval stages each followed by a moult. The infective stage for parasitic nematodes varies with species: it is often, but not invariably, the third stage larva.

To overcome the turgor pressure of the pseudocoelomic fluid the food, which is invariably fluid or contains minute particulate matter, e.g. bacteria in a fluid matrix, has to be pumped into the gut. The anterior region of the gut is thus a muscular **pharynx** (also, and perhaps inaccurately, known as the **oesophagus**) with a triradiate lumen which again differs in shape for different groups of nematodes. Behind the pharynx the gut continues posteriorly as the intestine. The intestinal wall is only one cell thick and the pseudocoelomic pressure means that the intestine is normally flattened unless there is food in it. The anterior intestine is secretory and releases digestive enzymes, while the more posterior intestine is absorptive. The intestinal lining cells have microvilli on the luminal side to increase the absorptive surface. There is no circulatory system and absorbed nutrients must cross the pseudocoelom to reach the muscles, the body wall and the reproductive system.

Posterior to the intestine is the cuticle-lined rectum. To prevent food which is pumped into the anterior of the gut passing straight through and out of the anus the junction of the intestine and rectum is marked by a sphincter muscle to control the passage of the faeces. The anal opening consists of a small slit which in most animal parasitic nematodes is sub-terminal. By definition the body between the anal opening and the posterior extremity is called the tail. In male nematodes the intestine and the reproductive system open into a cuticle-lined **cloaca** which also contains the secondary male sexual structures.

Although there are a few nematode species that are **hermaphrodite** or **parthenogenetic** most have separate sexes. Nematode reproductive systems are also tubular. The male system may be either single or double. The testis passes posteriorly *via* a seminal vesicle to the vas deferens, the distal portion of which constitutes an ejaculatory duct that opens into the cloaca. Most male nematodes also possess a pair of protrusible cuticular **spicules** which hold open the vulva and vagina of the female during copulation: these are often very distinctive. The posterior cuticle of the male may also be modified to provide other secondary sexual structures including suckers or a **copulatory bursa** supported by cuticular rays.

Most female nematodes have paired ovaries where oogenesis occurs. The developing eggs then pass through the oviducts and are fertilised by amoeboid sperm that is stored in seminal receptacles at the junction of the oviducts and uteri. These lead, often *via* a muscular **ovijector,** to a short vagina which opens through a slit-like vulva. The position of the vulva and the disposition of the reproductive organs varies from species to species.

The great majority of nematode eggs are ovoid with thin, three-layered shells. Most lack an operculum. Despite the enormous size range of adult female nematodes their eggs vary little in size, most are in the range 70–150 μm$\times$50–90 μm. Large female nematodes produce more eggs rather than larger ones. *Ascaris* females, for example, lay up to 200,000 eggs per day. Inevitably amongst the vast range of nematodes there are exceptions; a few produce eggs up to 350 μm long, some lay operculate eggs, and some provide their eggs with extra egg shell layers which may be distinctly sculptured. In a few species the eggs hatch in the uterus and first stage larvae are released from the female. The females of one important group of animal parasites, the filarial nematodes, which include such important species as *Wuchereria bancrofti* and *Onchocerca volvulus* that live in the tissues of their hosts, produce immature first stage larvae known as **microfilariae** which are small enough to be ingested by biting fly intermediate hosts.

The basic nematode life-cycle has six stages: the adults, the eggs and four larval stages. The larvae are still recognisably nematodes but lack the sexual organs and other adult features. The second, third and fourth larval stages and the adults are preceded by a cuticular moult when all the cuticular structures including the lining of the buccal cavity and the rectum are replaced. As there is no metamorphosis it has been argued that the immature nematode stages should be called juveniles rather than larvae and indeed many nematologists who study plant parasitic nematodes use the term juvenile. However, the use of larva is deeply entrenched amongst animal parasitologists and will be used here.

Thus, if our rumour-mongers had taken the trouble to look closely they would have found size differences, modified oral and pharyngeal structures associated with different feeding modes, males with distinctive spicules and copulatory structures and females with characteristic positioning of the vulva and reproductive organs often producing recognisably different eggs or larvae, with which to distinguish different nematode species.

Acanthocephala

The Acanthocephala or spiny-headed worms (Fig. 2.11) constitute a small phylum of exclusively parasitic helminths. Allegiances have been suggested with the nematodes – both are pseudocoelomate, round in cross-section and have separate sexes – and the cestodes, as both lack a gut and have an anterior attachment organ. However, they have a number of unique morphological and biological features and they are best accorded phylum status.

The most distinctive feature is the anterior **spiny proboscis** (see also Fig. 5.8) used to anchor the parasite to the intestinal wall of its host. It is covered with backward-facing spines, the number, shape and arrangement of which may be of diagnostic value. The proboscis in many acanthocephalans is retractable by muscular contraction into a **proboscis receptacle** and is everted by hydrostatic pressure in the receptacle wall. The body wall, like that of cestodes, is both protective and absorptive, and is complex and metabolically active (see Fig. 5.12). It contains within it a series of fluid filled **lacunar canals** which link with **lemnisci** which grow from the base of the neck. The exact function of these structures remains controversial but it is thought that the lemnisci and anterior lacunar canals may contribute to the hydraulics of eversion of the proboscis, while the more posterior lacunar system may act in part as a hydrostatic skeleton and in part as a rudimentary circulatory system transporting nutrients directly to the muscles of the body wall.

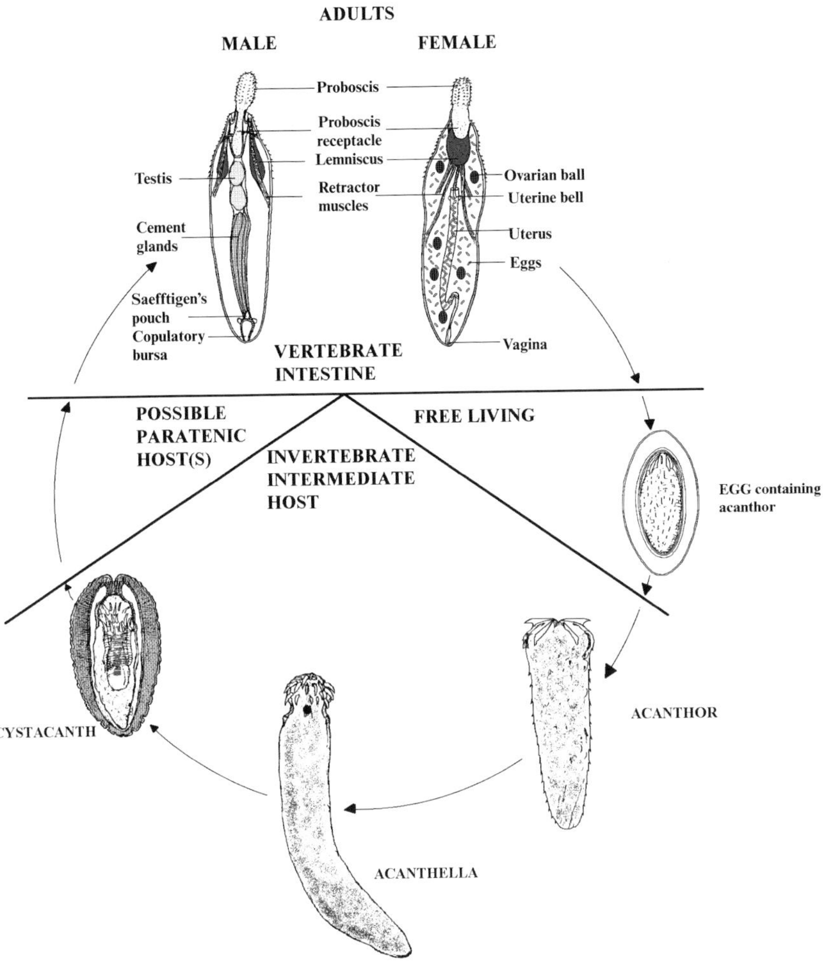

Fig. 2.11 General acanthocephalan life-cycle. The eggs, which are fully embryonated when laid, are ingested by an invertebrate intermediate host. They hatch in the intestine and the acanthor enters the body cavity, develops through the acanthella stage and encysts as a cystacanth. The definitive host may be infected by directly ingesting the intermediate host or a paratenic host which has eaten the intermediate host.

The lack of a functional gut restricts the Acanthocephala, like the cestodes, to the gastro-intestinal tracts of their hosts, as they rely on the host's digestive processes to perform the initial stages of food breakdown.

The reproductive systems develop in thin-walled **ligament sacs** within the pseudocoelom. The male system consists of two testes and associated sperm ducts which lead to a penis that opens through an evertible posterior bursa. The bursa is extended by the hydrostatic pressure developed in the associated **Saefftigen's pouch**. Also opening into the bursa are the ducts from a number of **cement glands** which, as we will see later, play a role in acanthocephalan mating biology.

The females are generally larger than their respective males and have a reproductive system that is superficially simple. The ovary breaks down during embryological development and **ovarian fragments** circulate freely in the ligament sac. These produce **oocytes** on their surface that eventually develop into eggs. The simplicity is deceptive as there is a complex filtering system that ensures that only mature, fertilised eggs exit through the posterior vagina. The selection is performed by the muscular **uterine bell.** Smaller immature eggs can pass through the maze of the uterine bell back into the ligament sac, while larger mature eggs pass through the vagina and hence to the lumen of the host's gut.

The lack of many of the usual female reproductive structures poses problems to which the acanthocephalans have evolved unique solutions. The male copulatory bursa is everted around the posterior of the female, the penis is inserted into the vagina and insemination occurs. There is no seminal vesicle to store the sperm and the vagina and uterus do not close tightly enough to prevent the sperm escaping from the female. To overcome this problem the secretions from the male cement glands provide a cap over the vaginal opening. This hardens to provide what in effect is a post-copulatory chastity belt. Although only temporary – the copulatory cap breaks down in a week or so – it does ensure that the female does not mate immediately with another male and that the outer oocytes of the ovarian fragments are fertilised, thus safeguarding the male's genes.

Acanthocephalans have indirect life-cycles with arthropod intermediate hosts which, depending on the species, may be terrestrial, fresh water or marine. The eggs when released from the host are immediately infective to the intermediate host and contain a mature **acanthor**, the first larval stage. On ingestion by the arthropod the acanthor emerges, penetrates the gut wall and enters the haemocoel where it elongates and grows through an intermediate stage known as an **acanthella.** Development of most of the adult organs is complete in the acanthella but it is not infective until it has rounded up and become surrounded by a membranous envelope. At this stage it is known as a **cystacanth**. The cystacanth remains dormant in the tissues of the arthropod until it is ingested by the definitive host where it excysts, everts the proboscis and attaches to the intestinal wall. If the arthropod is ingested by an unsuitable vertebrate the cystacanth may emerge and re-encyst in the tissues of what then becomes a paratenic host. Transmission occurs when this latter host is eaten by the definitive host.

Arthropoda

The phylum Arthropoda contains an enormous collection of species which, like the nematodes, occupy virtually all available ecological niches. There are more named species of insect than all other animals put together. Unlike the nematodes arthropods show no morphological constancy and the only features that all share are a chitinous exoskeleton and jointed appendages. The taxonomy of the arthropods is contentious. Fundamental questions about whether they have **monophyletic** or **polyphyletic** origins and the relationships between the various taxa are regularly raised. For our purposes it is probably best to consider them as members of a single phylum with each of the major divisions accorded subphylum status (Table 2.1).

Parasitism has been adopted as a way of life by many arthropods. Depending on how broad a definition of the term parasite is accepted different authors have estimated that between 15% and 50% of known insect species are parasitoids or have parasitic stages in their life-cycles. Some groups have a greater proportion of economically important parasites than others. Amongst the Chelicerata the ticks and many of the mites are important parasites. The Crustacea contains some of the most morphologically adapted of all parasites, e.g. *Sacculina* and *Peltogaster* (Fig. 2.12), whose allegiances have only been resolved by examination of their larval stages, as well as the more representative sea lice. The Hexapoda (Insecta) contains the obligate ectoparasitic fleas and lice as well as the blood feeding flies that act as vectors for other diseases.

Traditionally the parasitic arthropods have been studied by several different groups of zoologists. The Crustacea have generally been the province of fresh-water and marine biologists while the terrestrial arthropods have been studied by entomologists, even though the animals are not always insects and journals devoted to medical and veterinary entomology contain papers on ticks and mites as well as insects. In addition specialist acarologists who study ticks and mites have their own journals.

The life-cycles of arthropods fall into two broad categories (Fig. 2.13). In a **hemimetabolous** life-cycle the immature stages bear a superficial resemblance to the adults and there are only gradual changes between instars until the sexually mature adults develop. A **holometabolous** life-cycle includes a full metamorphosis with a larval stage adapted for feeding, a transitional pupal

Table 2.1. *Outline classification of the Arthropoda. Some authors see the arthropods as a polyphyletic grouping and in this case all taxa move up one level. Groups containing a large proportion of parasites and important parasites are indicated in bold type*

Subphylum	Class	Subclass	Order	Examples
Chelicerata	Merostomata			Horseshoe crabs
	Pycnogonida			Sea spiders
	Arachnida		Araneae	Spiders
			Opiliones	Harvestmen
		Acari	**Metastigmata (Ixodida)**	**Ticks**
			Prostigmata (Trombidiformes)	Harvest mites
			Mesostigmata (Gamasida)	Red mites of birds and mammals
			Cryptostigmata (Oribatida)	Beetle mites
			Astigmata (Sarcoptiformes)	**Mange mites**
Crustacea	Ostracoda			Ostracods
	Copepoda		Cyclopoida	*Cyclops, Lernaeocera*
	Branchiura		**Arguloida**	***Argulus***
	Cirripedia		Thoracica	Barnacles
			Rhizocephala	***Sacculina, Peltogaster***
	Malacostraca		Isopoda	Woodlice, fish lice
			Amphipoda	*Gammarus* Whale lice
			Decapoda	Crabs, lobsters, *Pinnotheres*
Uniramia	Myriapoda		Chilopoda	Centipedes
			Diplopoda	Millipedes
	Hexapoda (Insecta)		**Mallophaga**	**Chewing lice**
			Anoplura	**Sucking lice**
			Siphunculata	**Fleas**
			Diptera	**Flies**
			Hymenoptera	**Bees, wasps, ants (up to 50%**

Table 2.1 (*cont.*)

Subphylum	Class	Subclass	Order	Examples
				of species may be parasitoids)
			Coleoptera	Beetles
			Lepidoptera	Butterflies and moths
			+25 other orders	

stage and an adult reproductive stage. The different stages show no morphological similarity.

Crustacea

The crustaceans are largely aquatic as their exoskeletons are permeable to water, and any terrestrial crustaceans, e.g. woodlice, are restricted to areas of high humidity. Lobsters, crabs and prawns are familiar to most people because of their gastronomic connections but their parasitic relatives are less well known.

The crustacean body is typically segmented with a pair of appendages on each segment. The head characteristically consists of six segments with five pairs of appendages including two pairs of antennae, a feature which distinguishes the crustaceans from other arthropods. The remainder of the body generally consists of a thorax and an abdomen; the number of segments and appendages on each of these varies for each group.

Many of the crustacean classes contain parasitic examples, while a few are exclusively parasitic or contain a large proportion of parasitic species. Some, e.g. the ectoparasitic branchiurans, copepods and amphipods *Argulus, Ergasilus* and *Cyanus* are recognisably crustaceans, while the adult forms of others are so adapted that their origins can only be determined by examining their larval stages. The adult stages of the parasitic copepod *Lernaeocera branchialis* (Fig. 2.14) from the gills of cod and whiting, and the parasitic barnacle *Sacculina carcini* from the abdomen of the shore crab *Carcinus maenas*, consist of root-like attachment organs which may ramify throughout the tissues of their host, and a sac-like body containing the reproductive organs. Both species lay eggs which hatch to release typical nauplii which develop to copepodid and cypris larval stages respectively before attaching to a host.

Fig. 2.12 The parasitic barnacle *Peltogaster* on the abdomen of the hermit crab *Eupagurus bernhardus*. As an adult *Peltogaster* has few crustacean characteristics and its true relationships were only confirmed by examination of typical barnacle nauplius larvae.

As well as being parasitic themselves crustaceans may act as intermediate hosts for a range of parasites with aquatic stages in their life-cycles, e.g. the copepod *Cyclops* ingests larvae of the Guinea worm *Dracunculus medinensis* which after development are transmitted back to man when the *Cyclops* is itself taken in drinking water; *Cyclops* is also the first intermediate host for the pseudophyllidean tapeworm *Diphyllobothrium latum* (Fig. 2.9). Amphipods and isopods are common intermediate hosts for acanthocephalans.

Chelicerata

Members of subphylum Chelicerata differ from the other arthropods in that they lack antennae and have the first pair of appendages modified to form pincer-like **chelicerae**. The class Arachnida contains the free-living spiders, harvestmen and scorpions and also the parasitic ticks and mites which are included in a single subclass, the Acari.

Characteristically the body plan of arachnids consists of an anterior **prosoma** or **cephalothorax** which fulfils both the function of head and

HEMIMETABOLOUS
e.g. ticks, mites, lice

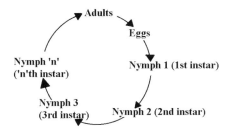

HOLOMETABOLOUS
e.g. fleas, flies

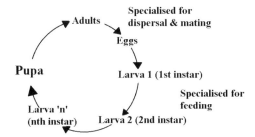

Fig. 2.13 Arthropods may have one of two basic life-cycles. In a hemimetabolous life-cycle the eggs hatch and a series of nymphal stages that more or less resemble the adults precedes maturation to the adults. Holometabolous life-cycles include complete metamorphosis, the larval stage is adapted for feeding and is followed by total reorganisation during the pupal stage and emergence of the adult stage adapted for dispersal and reproduction.

thorax and bears the appendages, and a posterior **opisthosoma** or abdomen which contains the respiratory and reproductive systems. In the spiders and scorpions the prosoma and opisthosoma are clearly delineated but in the ticks and mites they are fused to form a more or less globular body. The mouthparts vary from order to order but basically consist of a pair of **pedipalps** and a pair of chelicerae. The adults normally have four pairs of walking legs.

Acari – ticks and mites The subclass Acari (Fig. 2.15) contains a number of orders, one of which, the Metastigmata (Ixodida), contains the ticks, while

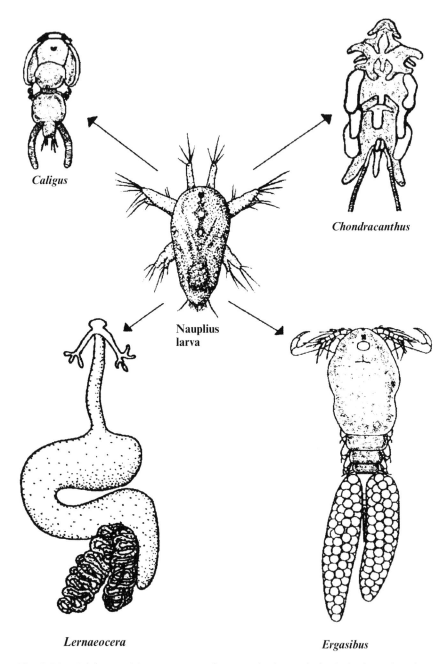

Caligus

Chondracanthus

**Nauplius
larva**

Lernaeocera

Ergasibus

Fig. 2.14 Adult parasitic crustaceans show marked morphological adaptations but all have a recognisable common nauplius larva testifying to their relationship.

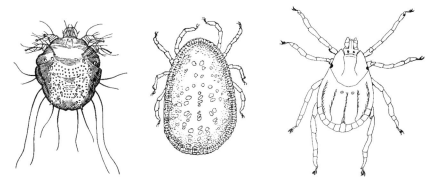

Fig. 2.15 Acari: a mange mite, a soft tick and a hard tick. All have a more or less globular body without external divisions. Ticks have a toothed hypostome lacking in mites. The hypostome is ventral in soft ticks but terminal in hard ticks, which also have a hard scutum on the dorsal surface. All have a hemimetabolous life-cycle.

the others contain the myriad mites. Mites are small – most are less than 1 mm long – extremely plentiful and widespread. Large numbers are free-living some of which, for example the oribatid mites, act as intermediate hosts for the cestode parasites *Anoplocephala* and *Moniezia* of horses and sheep respectively. The most important parasitic mites are those that cause mange, e.g. *Sarcoptes scabiei* that results in scabies in man and *Psoroptes ovis* that causes sheep scab in sheep. These are included in the order Astigmata (Sarcoptiformes).

While most mites are very small the ticks are larger: a fully engorged female cattle tick may reach 3 cm in length. Apart from size the most significant differences between mites and ticks relate to their mouthparts. Both groups possess chelicerae and pedipalps but the **hypostome** of ticks, which attach to their hosts and feed for extended periods, is covered by recurved teeth. There are two families of ticks, the Ixodidae or hard ticks and the Argasidae or soft ticks. The mouthparts of hard ticks are terminal and visible from above while those of soft ticks are ventral and thus not. The other main morphological difference between the two families is the possession by hard ticks of a sclerotised dorsal shield or **scutum** which covers the anterior of the dorsal surface on all stages except the adult males when it extends to cover the whole of the dorsal surface.

Both ticks and mites have hemimetabolous life-cycles with the pre-adult stages superficially resembling the adults. The larval stages which emerge from the eggs may however only have three pairs of legs, a feature which has caught out a number of unprepared and unsuspecting students under the stress of examination conditions who have regarded any six-legged arthropod as an insect! Hard ticks have a single nymphal stage but soft ticks and mites may have a number of such stages before the adult develops. Some ticks, one host ticks, spend their entire life on a single host individual before the fertil- ised female drops off to lay her eggs on the pasture. Others, two host ticks, spend their larval and nymphal stages on one host and the adult stage on a different one, while three host ticks attach and feed as larvae on one host, drop off and moult to nymphs before attaching and feeding on a second host and then dropping off and moulting to adults which attach and feed on a third host. While some ticks show a marked host preference most are more catholic in their tastes and two and three host ticks may attach to a variety of hosts. This behaviour makes ticks ideal transmitters of other pathogens, and, while their blood-sucking activities may do direct damage to their hosts, their role as vectors is probably more important. The wide host range means that ticks provide an easy route for pathogens to cross ecological boundaries.

Uniramia

There are two external morphological features that distinguish the uniramians from the other arthropods. Firstly none of their appendages are branched, they are all **uniramous**, hence the name, and secondly they have only one pair of antennae. The Uniramia includes the millipedes and centipedes but is dom- inated by the insects (Hexapoda).

The insect body is divided into three more or less clearly distinguished parts: the head which bears the antennae, the eyes and the mouthparts; the thorax with its three pairs of legs and in many cases one or two pairs of wings; and the abdomen which contains the digestive and reproductive organs. The range, diversity and number of insects is staggering; one author has calculated that there are 1×10^{18} in the world at any one time! Close to three-quarters of a million species have been described and these are distributed among some 30 orders. Some of these like the Coleoptera (beetles), the Hymenoptera (wasps, bees and ants) and the Lepidoptera (butterflies and moths) have hundreds of thousands of species while others, for example the Ephemeroptera (mayflies), Dermaptera (earwigs) and Anoplura (sucking lice) have only a few hundred. Parasitism in one guise or another occurs in many orders. In some it is rare – only about 60 of the 150,000 or so named lepidopterans are parasites or par- asitoids – whereas in other orders it is much more common. About half of the 200,000 hymenopteran species are parasitoids, and in yet other orders, e.g. Siphonaptera (fleas), all are parasitic. The Mallophaga (biting lice), Anoplura (sucking lice) and Siphonaptera (fleas), together with the Diptera (flies) of which about one tenth of the 100,000 species are parasitoids or ectoparasites

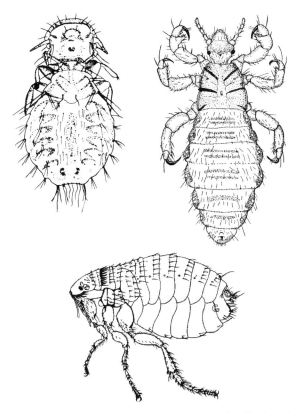

Fig. 2.16 Parasitic insects: a biting louse (Mallophaga), a sucking louse (Anoplura) and a flea (Siphonaptera). Lice are dorso-ventrally flattened and have a hemimetabolous life-cycle, while fleas are laterally flattened with many backward facing spines and are holometabolous.

and may act as vectors of other parasitic infections, normally exercise the attention of medical and veterinary parasitologists (entomologists).

Mallophaga and Anoplura – lice   The lice are small, dorso-ventrally flattened, wingless obligate parasites of birds and mammals. Of all the parasitic insects they are the most committed to parasitism, spending their entire life-cycle on their hosts. Transmission only occurs during nursing, mating or other close contact between individuals. It is rare for a louse not to be in close physical contact with a host, and absence from a host rapidly results in death. In man lousiness is associated with poverty and deprivation and especially the intimate associations imposed by warfare and its aftermath.

Lice (Fig. 2.16) are divided between two orders, the Mallophaga or biting lice and the Anoplura or sucking lice. Externally they may be distinguished by the fact that in the Mallophaga the head is broader than the thorax and the antennae have three segments, while the head of the Anoplura is narrower than the thorax and the antennae have five segments. The main reason that they are divided into two orders is that they have different mouthparts and feeding methods as well as rather different host ranges and habits. The Mallophaga are predominantly bird parasites although some live on mammals. They have biting mouthparts and generally feed by chewing feathers and hair, although some may take blood from lesions caused by the host scratching or from holes pierced into the feathers or skin of their hosts. The Anoplura are all mammal parasites and have skin-piercing mouthparts to feed on blood.

All lice are highly host and site specific and the entire hemimetabolous life-cycle is spent on the host. Eggs (nits) are cemented onto feathers or hair and rely on the body temperature of the host to develop and hatch. There are generally three nymphal stages before the adults and the whole life-cycle takes about a month. As it is the host's body temperature that governs reproduction and development lice do not show marked seasonality, and populations may increase rapidly – over 10,000 individuals have been reported from one shirt! Louse populations on wild animals however vary considerably in size and are often quite small.

From a human point of view lice are generally a nuisance rather than serious pathogens, but under conditions of close confinement and reduced hygiene they may achieve major pest status. Notoriously lice are vectors of typhus, caused by *Rickettsia prowazeki*, which affected 30 million people in eastern Europe in World War I, killing 3 million of them.

Siphonaptera – fleas   Like lice fleas are small, wingless insects ectoparasitic on birds and mammals. Rather than being dorso-ventrally flattened they are laterally flattened and highly adapted to living and moving rapidly through the dense jungle that is their host's coat. They are streamlined: the head, thorax and abdomen are fused and the antennae can fit into grooves on the head. Fleas are active animals and extremely slippery, as anyone who has tried to pick one up with their fingers can testify. The body is covered with backward facing spines and setae so the flea can move easily forwards but the host has great difficulty in pulling it backwards, thus increasing its chance of escaping.

Possibly the best-known feature of flea biology is their almost legendary ability to jump. They lack wings, although wing buds have been reported in the pupae, and the absence of wings in the adults is seen as an adaptation to parasitism. The loss of flight has been compensated for by the possession of a specialised jumping mechanism. The muscular energy in the legs is not directed immediately to jumping but is stored in a special pad made of **resilin**. Resilin is a protein that, like rubber, can store and release energy. The energy generated by initial contraction of the hind leg muscles is stored in the resilin pad and then, as the leg continues to move, a catch-like mechanism operates to release all the energy in a single explosive burst. This can launch the flea up to 30 cm horizontally and 20 cm vertically with an acceleration exceeding 130 g. The equivalent for man would be a long-jump of 130 m and a high-jump of 85 m! The front legs are generally held above the back of the flea, and as it jumps it rotates in the air. The claws on the ends of the legs can then act as grappling hooks to aid attachment to the hair of any host that is encountered.

Many fleas, unlike lice, spend much of their time away from their hosts in the hosts' nests, lairs or burrows. They are also less host specific than lice and, if hungry, will feed on a range of available hosts. This phenomenon is experienced by people who move into a house that has stood empty for some time after the previous occupants had kept a dog. The fleas, having been starved, leap gratefully on anyone who comes into range and can provide an unwelcome house warming party. Not all hosts may be equally suitable for completing the life-cycle and many fleas do show a considerable adaptation to a particular host.

Fleas have a holometabolous life-cycle. The eggs are laid either in the nest or on the coat of the host and are then dislodged when it settles down. The legless, eyeless maggot-like larvae hatch and feed on organic debris in the nest before spinning a silken cocoon and pupating. The period of pupation is often used to overcome periods of adverse weather conditions and to allow overwintering away from a seasonally migrating host. The newly emerged adults are very sensitive to the presence of potential hosts and different species have been shown to respond to vibration, temperature change, exhaled carbon dioxide, shadows and specific odours.

Fleas have a nuisance value to man but, like the lice, are much more significant as vectors. Notoriously they transmit *Yersinia pestis*, the bacterium responsible for causing bubonic plague, the Black Death of the 13th and 14th centuries which killed about a quarter of the population of Europe. Plague circulates in wild rodent populations, transmitted between them by the bite of fleas, e.g. the oriental rat flea *Xenopsylla cheopis*. The bacteria multiply rapidly in the gut of the flea to such an extent that the fore gut becomes blocked and in trying to take a further blood meal the flea regurgitates some of the bacteria into the bloodstream of its victim. Epidemics occur when domestic rats become infected, the rats die and the fleas transfer to other hosts including man. Today plague tends to be a focal disease but is still present in wild mammals in many parts of the world, and a breakdown in conditions, e.g. local wars, earthquakes, etc. may be all that is needed to initiate an epidemic.

Diptera – flies

Unlike other flying insects the flies have only one pair of wings. The second pair is modified into two club-shaped **halteres** which beat rapidly and function as gyroscopic organs, increasing stability in flight.

Fly mouthparts are typically adapted for fluid feeding. Some, like the houseflies and blowflies, have mouthparts which act like blotting paper, sucking up liquids through fine grooves by capillary action. The grooves are too narrow to accept most parasite transmission stages although some small eggs may be ingested and these flies may mechanically transmit them along with viruses and bacteria that adhere to the mouthparts or body. In this way *Entamoeba* cysts may be transported from faecal material to food preparation areas and effect transmission. Although the adults are unable to ingest parasite stages, their larvae can. Thus the nematode *Habronema* which lives in the stomach of horses produces eggs that are eaten by larvae of houseflies and stableflies that develop in manure. The eggs hatch in the gut of the maggot, penetrate the gut wall and mature to infective third stage larvae about the time of pupation. In the adult flies that emerge the nematodes pass to the mouthparts from which they are deposited on the lips of the horse when the fly feeds. Transmission may also occur if the whole fly is ingested, perhaps after falling into water and drowning.

Other flies, e.g. mosquitoes, tsetse flies and sandflies have mouthparts adapted for piercing and sucking. Those that feed on blood may act as vectors of blood and tissue viral, bacterial, protozoal and helminth diseases. In some species both males and females are blood feeders but in others, for example mosquitoes, only the females require a blood meal to produce their eggs. The males feed on nectar and other plant juices.

Flies have a holometabolous life-cycle. Most lay eggs but the site of egg deposition varies with species. Mosquitoes, blackflies and midges lay their eggs in water and have aquatic larval and pupal stages, while sandflies lay theirs in crevices between rocks or in walls and are terrestrial. Female tsetse flies are unusual as they retain one larva at a time within their body, nourished by secretions from special 'milk glands' until it is fully grown and ready to pupate. It is then deposited on loose, dry soil into which it can burrow and complete pupation.

Houseflies and blowflies deposit their eggs in, and their larvae feed on, decomposing organic matter including rotting vegetation, manure, faeces and decaying carcasses. It is not a very large step from feeding on carcasses to feeding on wounds on living animals and a number of fly genera have adopted such habits. This may sound unsavoury but surprisingly they have been used to man's benefit. Excretions from the larvae have anti-bacterial properties and, before the advent of antibiotics, the larvae were used to clean up wounds and remove diseased tissue. While they feed preferentially on dead tissue they may transfer to living tissue in its absence, which was hardly to the patient's advantage. Following the development of antibiotic resistance by many bacteria the use of sterile maggots under carefully controlled conditions for cleaning persistent wounds is being re-examined.

These species are opportunist parasites but there are others which are more directly parasitic. Adult female screw-worms (*Callitroga hominivorax*) actively seek out the scent of fresh blood, commonly caused by a tick bite, to lay their eggs in the lesion. The eggs hatch and the larvae feed on the living tissue, causing severe distress and possible death of the host.

Other species do not need an initial lesion and are obligate parasites. Warble flies (*Hypoderma* spp.) lay their eggs on the intact skin of cattle. The larvae that emerge penetrate into the skin and may spend several months migrating around the viscera before returning to the skin where they form cyst-like warbles which have small breathing holes leading from the larvae to the surface. Eventually the larvae emerge from the warble, drop to the ground and pupate. The eggs of *Oestrus ovis* hatch in the female and the newly hatched larvae are deposited in or on the nostrils of sheep, goats and deer. The larvae invade the sinuses and grow to 2–3 cm before migrating back to the nostrils to drop out and pupate. Female *Gasterophilus* attach their eggs to the hairs of horses from which they are licked when the horse grooms. The eggs stick to the tongue, hatch and invade the tissues of the tongue where they spend 3–4 weeks before migrating to the stomach where they mature. When ready to pupate they detach from the stomach wall and pass through the intestine and out with the droppings.

Invasion by fly larvae is called **myiasis** and, while these species essentially cause problems in animals, most have been reported as occasional parasites of man; a prospect on which nightmares may be based.

3

From host to host

The most fundamental task that faces all parasites is that of locating another suitable host for their progeny, especially when many of them have high host specificity and thus a reduced host range. For a parasite established within a host that host constitutes its whole world and to launch its offspring into the unknown extra-host environment is somewhat akin to man launching into space exploration. The parasite, like the astronaut, must be protected against the vagaries of the climate and be able to survive for possibly an extended time in a hostile and alien environment. Transmission is a high risk period for any parasite and many produce vast numbers of transmission stages to overcome the randomness of the process; these seem to have more in common with the disordered wanderings of Red Dwarf than the planned precision of NASA. Other parasites have evolved various adaptations to increase the chances of successful transmission. Some of these are highly ingenious and sophisticated but even those parasites that have scorned sophistication may be very successful in terms of numbers of hosts infected, even if the chance of any single individual stage achieving its objective is extremely small. Some of the transmission problems appear almost insurmountable but are successfully solved. For example, many monogenean oncomiracidia have to locate and attach to the gills of a fish. The maximum speed for a 250 μm oncomiracidium is about 5 mm/sec, while a 10–25 cm fish host can swim at about 300–1500 mm/sec, i.e. 60–300 times the speed of the miracidium. In relative speed terms this is the equivalent of our trying to catch an aircraft as it flies by! Just as we go to airports where aircraft are easily boardable so the oncomiracidia show behavioural adaptations to bring them into areas where fish congregate. Some species only hatch when it gets light, an adaptation to

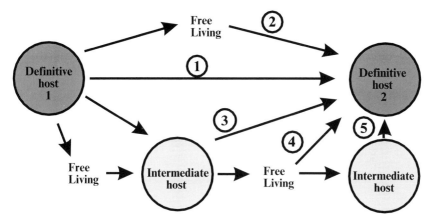

Fig. 3.1 Routes of parasite transmission from host to host. 1. Direct transmission, *e.g. Trichomonas* spp., transplacental and transmammary transmission. 2. Direct life-cycle, *e.g. Entamoeba, Giardia,* Monogenea, many nematodes. 3. Indirect life-cycle, with vector transmission *e.g. Trypanosoma, Plasmodium,* filarial nematodes. 4. Indirect life-cycle, *e.g.* many digeneans, cyclophyllidean cestodes, Acanthocephala. 5. Indirect life-cycle with two intermediate hosts, *e.g. Dicrocoelium,* pseudophyllidean cestodes.

allow them to gain access to night-feeding fish that rest during the day; some show positive phototaxis and negative geotaxis to take them towards surface-feeding fish, while others remain near the bottom to attach to bottom-dwelling fish; some show awareness of a potential host by recognising specific chemicals emanating from it and some respond to the water currents that flow around a swimming fish. These various reactions give the parasites the advantage they need to achieve infection. Each different parasite has its own problems to overcome and the solutions are many and various (Fig. 3.1).

Adaptations that facilitate direct transmission

We have already seen that *Trichomonas* spp. do not produce special transmission stages but pass directly from host to host by intimate contact. *T. tenax* passes orally and rarely if ever causes pathological changes. *T. vaginalis* on the other hand inhabits the urinogenital tracts, is transmitted venereally and in different surveys has been reported in 20–40% of women and 4–15% of men. In women it may cause clinical vaginitis although it rarely causes symptoms in men. Trichomoniasis is not restricted to the sexually active and transmission probably also occurs as a result of poor personal hygiene and contact with contaminated articles such as shared wash cloths. A related species *T. foetus* lives in the reproductive tract of cattle and may be responsible for causing abortion. *T. gallinae* lives in the crop of birds, especially pigeons, and is passed directly from the mother to her chicks when she feeds them the 'pigeon milk' which she produces for them in her crop. Some strains of the parasite are highly pathogenic and may kill the chicks; any that recover are resistant to further infection but remain symptomless carriers to infect another generation.

Transmission of parasites from mother to offspring is highly efficient. Not only are the young available and vulnerable, they may be more susceptible as their defence mechanisms may not be fully developed.

Toxocara canis is a relatively large (up to 15 cm) nematode parasite from the intestine of dogs (Fig. 3.2). The adult females produce eggs with a thick protective shell that are passed in the faeces. They embryonate in the environment over several weeks depending on temperature until they contain infective, second stage larvae. If these infective eggs are ingested by a pup up to about six months old they hatch in the intestine and go on a circular tour of the pup's body. Initially they penetrate the intestinal wall and pass in the blood stream firstly to the liver and then to the lungs, where they break into the alveoli and are carried by ciliary action up the bronchial tree to the oesophagus. They are then swallowed, pass through the stomach and re-enter the intestine where they mature to adults and start to lay eggs. The precise reason why the larvae need to undergo this so-called **tracheal migration** to reach the point from which they started is not known. One explanation is that these present-day parasites of terrestrial hosts evolved from ancestors in aquatic hosts that had intermediate hosts in their life-cycles. Such parasites still exist in marine mammals and the larval stages can sometimes be seen in the flesh of cod or whiting. The suggestion is that in the transition from the sea to land the parasites lost their intermediate hosts but not the need to migrate.

If embryonated *Toxocara* eggs are eaten by older dogs they hatch and the second stage larvae that emerge penetrate the gut mucosa as usual but, instead of completing the tracheal migration and returning to the intestine, they undergo what is known as **somatic migration**, enter the tissues and become dormant. In dogs this tends to be a dead end but in bitches the larvae are triggered by hormonal changes during the later stages of pregnancy to move to the placenta and mammary glands. Those larvae that go to the placenta cross it and enter the unborn pups, where they congregate in the liver and pass to the lungs at the time of birth. They then migrate to the intestine *via* the tracheal route and mature to adults within about a month of birth. Larvae that migrated to the mammary glands are passed to the pups from the nursing

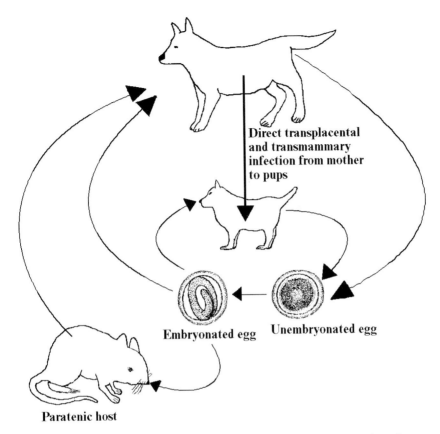

Direct transplacental
and transmammary
infection from mother
to pups

Embryonated egg   Unembryonated egg

Paratenic host

Fig. 3.2 *Toxocara canis* life-cycle. Transmission may be by direct ingestion of embryonated eggs containing second stage larvae or by predation on paratenic hosts that have eaten the eggs. Most transmission however occurs directly from bitch to pup by transplacental or transmammary infection.

bitch, so there is the opportunity of a second dose. **Transplacental** and **transmammary infection** mean that a very large proportion, over 90% in some surveys, of pups are infected with *Toxocara*. Adult dogs more rarely have adult worms as most of the larvae that enter them pass into the tissues; in this way an individual may build up a large population of larvae from a number of small doses. Not all larvae are mobilised at once so a bitch may infect more than one of her litters even if she is kept worm-free between whelpings.

If the embryonated infective eggs are eaten by another animal – mouse, rat, etc. – they hatch and somatic migration occurs, with the larvae lying dormant in the tissues of what has now become a paratenic or transport host. The larvae remain viable and can be transmitted when the mouse or rat is preyed upon. This is probably not as important a route of infection for *T. canis* as it is for the related species *T. cati* which, as the name implies, lives in cats. Cats are renowned predators and small mammals probably constitute a greater part of their diet than they do for most domestic dogs. *T. cati* does not undergo transplacental migration and kittens born by Caesarean section are not infected, but transmammary infection does occur and again a large proportion of kittens become infected from their mother's milk.

The range of potential paratenic hosts is very wide and includes man. The migrations of the larvae, generally believed to be *T. canis* but *T. cati* may also be involved, may result in man in a condition known as **visceral larva migrans**. This is marked by a number of symptoms including wheezing, anaemia, abdominal pain, swollen liver and swollen spleen, and some 300 cases are diagnosed each year in the UK. Some of the larvae may enter the eye where they may cause irreparable damage to the retina. In the past some of these lesions were mistakenly diagnosed as being malignant and resulted in the removal of the eye. Fortunately awareness of the problem coupled with modern diagnostic and treatment methods makes such disasters much less likely. It has been estimated that in Britain dogs produce some 1,000,000 kg of faeces each day containing 150,000,000 *Toxocara* eggs. Taking the average life span of the eggs as two years this means that there may be as many as 110,000,000,000 eggs in the environment at any time. A number to conjure with the next time you have a face-full of mud while playing rugby in a public park!

Synchronising host and parasite life-cycles

T. canis and *T. cati* larvae pass directly from mother to offspring but other parasites adjust their reproductive periods to those of their hosts to ensure that there is a ready source of susceptible hosts available.

The rabbit flea, *Spilopsyllus cuniculi*, unlike many other fleas which are essentially nest dwellers, spends much of its time attached by its mouthparts to the rabbit's ears, feeding on blood. Transfer from rabbit to rabbit occurs through contact and may be especially common at mating. Copulation causes excitement not only for the rabbits but also for their fleas: there is a marked rise in temperature and transfer between buck and doe may occur. Only the fleas that eventually remain on the doe breed. The hormonal changes that occur early in the doe's pregnancy cause the fleas to attach even more tightly and any extra fleas that may subsequently transfer to a pregnant doe also attach firmly.

About 10 days before the birth of the young rabbits there is an increase in the level of circulating corticosteroid hormones in the doe's blood. The

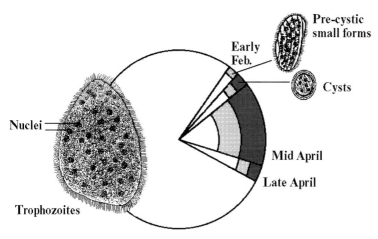

Fig. 3.3 Timing of *Opalina* life-cycle from the intestine of a frog. For much of the year *Opalina* exists as trophozoites in the lumen of the frog's intestine. Small forms and cysts are produced in response to hormonal changes in the frog so that the reproductive cycles of host and parasite are synchronised.

response of the fleas to this is for the eggs in the females to mature and for the guts of both the males and females to increase in size so that they ingest more blood, much of which passes straight through in their faeces and provides a source of iron for the subsequent flea larvae. At the time of parturition there are massive hormonal changes and the fleas detach from the ears and pass to the doe's face and thus to her young when she is tending them. The flea eggs by this time are mature and the fleas mate and lay their eggs. The adult fleas remain on the young rabbits while their circulating corticosteroid level remains high. As soon as it begins to drop, after about a week, they transfer back to the doe leaving their eggs to hatch and the larvae to feed on the dried blood diet supplement that has been provided for them.

Many amphibians live largely solitary lives and only congregate for short periods for breeding. Their parasites thus have special reasons to synchronise their life-cycles with that of their hosts.

The large, multinucleate protozoan *Opalina* is a common inhabitant of the intestine of frogs. For most of the year it exists as asexually reproducing trophozoites (Fig. 3.3). Early in the year small forms start to appear: these are pre-cystic stages and are followed by cysts that are passed in the faeces. The cysts are available to infect the tadpoles when they feed. Micro- and macro-gametocytes emerge from the cysts and give rise to uninucleate gametes which fuse to provide zygotes that grow into the trophozoites. The life-cycle of *Opalina*, like the breeding of the rabbit flea, is under hormonal control.

The monogenean *Polystoma integerrimum* lives in the bladder of frogs. During the winter the worms are inactive but in the spring, in response to changes in gonadotrophin levels in the host's blood, they mate and the fertilised eggs are passed in the urine. The development of the eggs takes about the same time as it takes for the tadpoles to reach the internal gill stage. The eggs hatch and the oncomiracidia enter the gill chamber of the tadpoles and attach to and feed on the gills. When the tadpole metamorphoses into a frog the young worms migrate at night over the ventral surface of the tadpole and enter the cloaca and bladder. The exact mechanism of control of the life-cycle has not been resolved but is probably hormonal, although experimentally a number of non-hormonal substances injected into non-breeding frogs can also initiate breeding changes in the parasites. Frogs do not reach breeding age for three years and nor do their *Polystoma*, which again suggests that there may be a close hormonal linkage between the host and its parasites.

The middle of the Arizona Desert is about the last place that one would expect to find a parasite that requires water for transmission in a host for which water is essential for reproduction. However that is where the monogenean *Pseudodiplorchis americanus* lives in the urinary bladder of the spadefoot toad *Scaphiopus couchii*. *Scaphiopus* spends up to nine months of each year hibernating buried up to a metre below the surface. During the brief summer rains the toads emerge to feed on the invertebrates that are plentiful at this time and to spawn in the temporary pools that form. They spend no more than about 24 hours a year in water. The parasites are adapted to synchronise their life-cycle with the toads. The oncomiracidia develop to maturity within the uterus of the worms and hatch immediately they are passed. They invade toads *via* the nostrils, migrate to the buccal capsule and thence to the lungs where they remain until stimulated to migrate through the digestive tract to the cloaca and bladder.

Synchrony of host and parasite life-cycles allows a parasite to maximise its reproductive effort, to ensure that a plentiful supply of susceptible hosts is readily available for its progeny. Such synchronisation may be controlled by the environment rather than the host.

Nematodirus battus is a nematode parasite from the intestine of lambs. It may cause an acute and fatal disease. It is unusual as it was only described as a separate species in the 1950s in East Anglia, and has since been carried to many other areas of the UK and also to Norway. It seems unlikely that such a damaging parasite would not have been recognised earlier and it has been suggested that it has only recently adapted to sheep, possibly from an exotic imported species, for example, muskrat or coypu. Transmission occurs from lamb to lamb; adult sheep are largely refractory to infection. The eggs that are

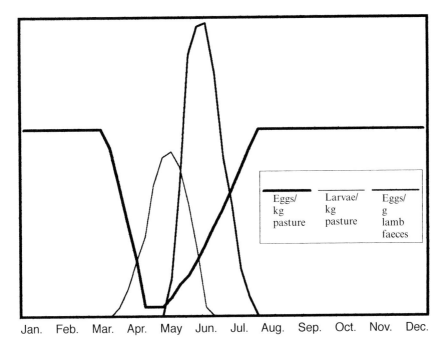

Jan. Feb. Mar. Apr. May Jun. Jul. Aug. Sep. Oct. Nov. Dec.

Fig. 3.4 Epidemiology of *Nematodirus battus* in Britain. Eggs produced in lambs build up on the pasture during the summer but do not hatch until the following spring, resulting in a sharp peak of infective larvae about the time the new lambs are grazing. The timing of the peaks and intensity of infection depend on weather conditions and vary from year to year.

produced by one season's lambs pass on to, and build up on the pasture (Fig. 3.4). They develop slowly through the summer and eventually reach the infective stage containing third stage larvae. They do not hatch but overwinter on the pasture and only hatch when the temperature reaches 10°C after a period of freezing. Once the mean daily temperature reaches 10°C the eggs hatch rapidly and result in a sharp peak of infective larvae on the pasture. If this peak coincides with the new season's lambs being on the pasture there is likely to be a large uptake of larvae and serious disease. In a warm, early spring the peak of larvae occurs while many of the lambs are still suckling and fewer ingest damaging numbers of larvae. The severity of nematodiriasis, the disease caused by *Nematodirus*, thus varies from year to year depending on the climate (see Chapter 8).

Many species of parasite apparently broadcast their progeny into the environment and rely entirely on random contamination to achieve contact with another host. As one looks more closely at these life-cycles more subtle

Table 3.1. *Size and daily egg production for a range of helminth species*

Species	Length of egg producing adult	Approx. eggs per day (unless otherwise stated)
Digenea		
Schistosoma mansoni	2 cm	500
Fasciola hepatica	3 cm	20,000
Cestoda		
Echinococcus granulosus	5 mm	100
Taenia hydatigena	5 mm	60,000
Hymenolepis diminuta	90 cm	250,000
Taenia saginata	5–10 cm	750,000
Diphylobothrium latum	10 cm	1,000,000
Acanthocephala		
Polymorphus minutus	1 cm	2,000
Macracanthorhynchus hirudinaceus	30 cm	260,000
Nematoda		
Trichinella spiralis	3 mm	1,500 during whole life
Enterobius vermicularis	1 cm	10,000 during whole life
Necator americanus	1 cm	5,000
Ancylostoma duodenale	1.5 cm	15,000
Trichuris trichiura	4 cm	5,000
Ascaris lumbricoides	30 cm	200,000

adaptations appear. Protozoan parasites multiply within their hosts so that theoretically a single ingested cyst can produce an infinite number of transmission stages. In practice this does not occur and a large number of limiting factors apply. Even so, large numbers of progeny are released. Helminths do not multiply in their hosts and their fecundity is largely limited by their size (Table 3.1). Parasites that rely on contamination and ingestion for transmission need to get into the food chain of potential hosts and adaptations to encourage this abound.

One of the commonest helminth parasites in man in Britain is the pinworm, *Enterobius vermicularis*, also sometimes called the seat worm or thread worm (Fig. 3.5). This relatively small nematode exploits our communal habits and commonly affects whole families or groups living together. The adults live in the large intestine. When the females are gravid and full of eggs, they

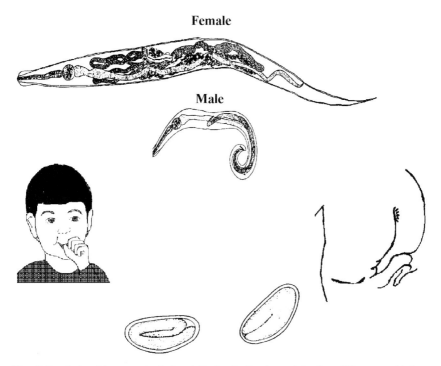

Fig. 3.5 *Enterobius vermicularis* adults live in the large intestine. When gravid the adult females migrate out of the anus and deposit their characteristically shaped eggs on the perianal skin. The irritation caused by the migrating worms often results in the host scratching and then transporting the eggs, which embryonate rapidly, back to the mouth.

migrate out of the intestine onto the skin around the anus at night when the body's core temperature is at its lowest. The air stimulates them to deposit their eggs and they then retreat back into the intestine. Migration onto the peri-anal skin causes irritation and, especially in children, the natural reaction is to scratch the affected area. Eggs may then become trapped under finger nails and, as they embryonate within a few hours, may well reinfect the same individual when the child sucks its thumb to soothe itself back to sleep. The eggs are very light so that when the bed clothes are shaken in the morning they are disseminated into the air to land on any surface and thus infect the rest of the family.

Enterobius eggs are not highly resistant to environmental conditions and die within a week or so. The same cannot be said for the eggs of *Ascaris*. This large nematode from the small intestine of man and pigs produces vast numbers of eggs. A single adult female may produce up to 200,000 eggs per day for up to a year and may contain 27,000,000 developing eggs. When passed in the host's faeces the eggs are unembryonated and only become infective when the larvae have developed to the second stage, still within the egg shell. The time taken for development depends largely on temperature. In the laboratory at the optimum temperature of 30°C it takes about three weeks for embryonation, but in temperate regions this is much more protracted under field conditions. Experimental work on the pig species *A. suum* has suggested that there is little development during the cooler months so there is a build-up of unembryonated eggs which all start to develop together when the temperature rises in the spring. The time for complete embryonation decreases as the daily mean temperature rises, and the result of this '**temperature telescoping**' is a peak of infective eggs available during the summer. Conveniently this coincides with the arrival of the new crop of piglets and helps to ensure transmission. The eggs are highly resistant to desiccation and freezing. One intrepid group of workers fertilised a patch of strawberries with untreated human sewage containing *Ascaris* eggs. Each year for the following six years they ate unwashed strawberries from the patch and each year received a small dose of *Ascaris*. Whether the eggs could have survived longer and had just been diluted and washed into the soil was not ascertained and, perhaps not surprisingly, the experiment has not been repeated! Laboratory studies have however demonstrated that the eggs are highly resistant to many disinfectants and other chemicals, and many unembryonated eggs permanently 'fixed' and mounted on microscope slides have continued to develop and when re-examined need to have their labels changed to embryonated.

The human hookworms *Ancylostoma duodenale* and *Necator americanus* (Fig. 3.6) affect some 1000 million people throughout tropical and sub-tropical regions. The adults live in the small intestine attached to the gut mucosa, sucking blood. The females release eggs that pass out, develop to first stage larvae and feed on bacteria in the faecal mass, before moulting to second stage larvae which continue to feed until moulting to the third stage. This second moult is incomplete and the third stage larvae retain the second stage cuticle around them as a **sheath** which may provide additional protection from desiccation. The infective third stage larvae have a non-functional pharynx and move away from the faeces. They infect another person by direct penetration through the skin, often the soft skin between the toes when the larvae are trodden on by bare feet. The larvae enter the blood stream and are carried in the blood stream to the liver and the lungs and then *via* the tracheal route to the intestine. In tropical areas transmission continues throughout the year but in areas of the world where there are marked wet and dry seasons transmis-

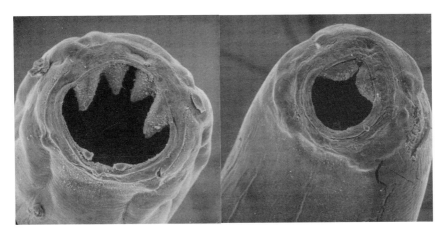

Fig. 3.6 Anterior of the hookworms *Ancylostoma duodenale* and *Necator americanus*. The adults suck a plug of gut mucosa into their buccal capsules and feed on the blood that escapes from the lacerations made by the teeth and cutting plates respectively.

sion may be prevented by the low humidity during the dry season. To overcome this *A. duodenale* larvae that develop in the soil at the end of the wet season as the faeces dry out do not complete development in their new host. They enter the host's tissues and remain dormant until the return of the rains when they complete their migration to the intestine and develop to adults which produce eggs that have an improved chance of suitable development conditions. While *Ascaris* provides its progeny with a highly resistant capsule, the egg shell, to avoid unfavourable conditions, *Ancylostoma* achieves the same result by protecting itself within its hosts. So far this **arrested development** has not been demonstrated in *Necator* infections.

Today human hookworms do not occur in the UK although they were common in mine workers. The stable, warm, damp and insanitary conditions of 19th century mines provided ideal conditions for hookworm transmission. There are however a number of more or less closely related species found in domesticated animals. These do not normally infect by skin penetration but by ingestion of infective third stage larvae. Many of these species exhibit a similar seasonality to hookworms, but the stimulus for initiating arrested development is decreasing temperature in the autumn rather than reduced moisture, although in different parts of the world with different climates moisture may again be a factor.

While seasonal climate changes vary in a broadly predictable way to provide cues for the developing larvae to enter **hypobiosis**, the internal environment of the host is more stable and a number of factors contribute to indicate the end of diapause. Firstly the larvae appear to have an in-built 'clock' so that after a predetermined time they complete their migration. Larvae experimentally induced into hypobiosis emerge into the intestine and complete development to adults regardless of the time of year the stimulus was applied. The clock is not necessarily very accurate nor is induction complete, so that emergence may occur over a broad time band but with a more or less marked peak. Changes in the host may also affect arrested larvae. Many animals are at their least healthy at the end of winter. The stresses of the environment coupled with a reduction in both the quality and quantity of food and often the pressures of pregnancy all contribute to a loss of condition. During the later stages of pregnancy and through lactation considerable hormonal and immunological changes occur in the mother, resulting in a lessening of immunological resistance. These changes also contribute to the maturation of arrested larvae. Additionally the reduction in resistance results in an increase in egg production by any residual females from the previous summer's intake. The net result of all of these is a marked increase in egg production from adult hosts (Fig. 3.7). As this occurs in the spring around the time that young animals are born it is sometimes known as the **'spring rise'** or **'periparturient rise'.** The peak of egg production in April–May results in a peak of infective larvae on the pasture in mid-summer and a consequent peak of eggs coming from the lambs or calves towards the end of the summer or early autumn. By utilising arrested development and exploiting the host's weakened immune response the parasites again ensure that their progeny are available on the pasture when the greatest number of vulnerable, young hosts are available.

One of the more unusual means of transfer is to rely on another parasite to provide protection. *Histomonas meleagridis* is an intestinal protozoan parasite of gallinaceous birds which, like *Naegleria*, can exist in both amoeboid and flagellate forms. In chickens it is a harmless lumen dweller but in turkeys it invades the gut mucosa and liver and may cause a fatal disease known as **black head**. The trophozoites do not form cysts and cannot survive passage through the acid in the anterior gastro-intestinal tract and for a long time the mode of transfer was a mystery. Electron microscope studies of both the adult and egg of the nematode *Heterakis gallinarum* surprisingly unravelled the story. *Histomonas* trophozoites are ingested by *Heterakis*. They cross the nematode's intestinal wall and pass through the pseudocoelom and enter the reproductive system. In the females they pass into the oviduct and become incorporated with the developing eggs inside the egg shell. The protozoan does not affect embryonation of the nematode larva and remains dormant until the egg is eaten by another bird. The egg shell provides protection from

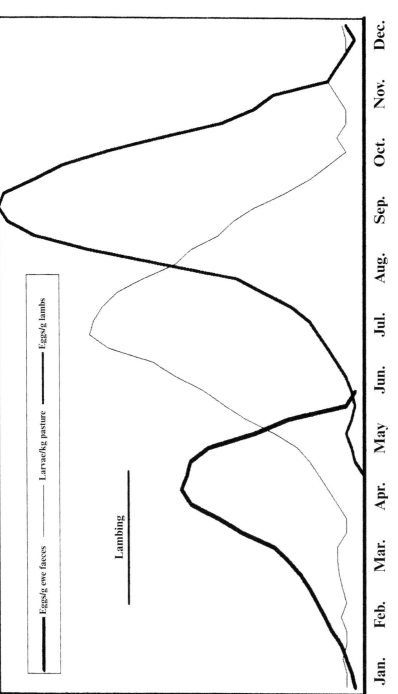

Fig. 3.7 Epidemiology of *Ostertagia* in sheep. Eggs released from ewes in the spring result in larvae on the pasture during the summer and infection of that season's lambs in the autumn. Larvae that hatch towards the end of the summer do not complete maturation but remain dormant in the host's tissues until the following spring, increasing the size of the spring peak.

the acid in the proventriculus and, when it hatches in the intestine, both the nematode and the protozoan are released to establish a new infection.

The small size of protozoans means that they can exploit such transmission routes. The mammalian blood sporozoan *Babesia* is transmitted by hard ticks. Ticks normally leave one host and moult before attaching to the next host so that the parasite has to be able to pass from one developmental stage to the next, **transtadial transmission**. In addition they can enter the eggs and pass from one generation of ticks to the next, **transovarian transmission**.

At the start of this chapter it was pointed out that the restricted host range of many parasites contributes to the problems of locating a suitable host. One way of easing this problem is to increase the host range. Few parasites have adopted this approach but one, *Trichinella spiralis*, exploits it to the full. *Trichinella* is a small (3–4 mm) nematode with a very broad geographical and host range. It extends from polar regions to the tropics in most mammals and many birds. Until fairly recently it was thought to be a single species, but up to four species are now more or less widely recognised. All species have a similar life-cycle and transmission pattern and are largely distinguished by their infectivity to different host species. The adults live and mate in the mucosa of the small intestine and the fertilised females deposit first stage larvae directly into the lymphatic vessels. The larvae, which are only about 100 μm long, pass from the lymph to the blood stream and are distributed around the body. They develop further only when they have invaded striated muscle cells, especially those of the diaphragm, tongue and cheek muscles. The muscle cell transforms into a nurse cell for the larvae which grow for about a month until 800–1000 μm long, when they coil up and encyst, still nourished by the modified host cell (Fig. 3.8). The larvae are now infective for another host which becomes infected when it eats the muscle. The presence of the larvae causes muscle dysfunction, especially in the respiratory system, and heavy infections may result in death. Lesser infections are likely to make the host less fit to escape a predator, into which the parasite can readily pass.

Man normally becomes infected from eating undercooked pork or pork products, the pigs having been infected from eating rats that have fed on discarded pork scraps, from cannibalism in overcrowded sties or from being fed undercooked swill. Over recent years there have been education campaigns in western nations to increase awareness of the problem and to reduce the level of infection. In the USA trichinosis was designated a reportable condition in 1947 and since then the number of cases reported has dropped from an annual average of 400 in the 1940s to 32 in 1994. Many of the cases now reported have had their origins in unusual dietary habits. One outbreak in Alaska was traced back to a single family party where grizzly bear meat was

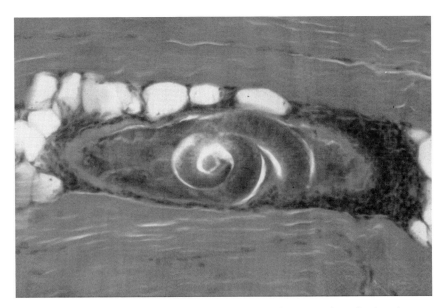

Fig. 3.8 Encysted first stage larva of *Trichinella spiralis* in an infected striated muscle cell. *Trichinella* is unusual as it encysts in its definitive host and relies on predation for transfer to a new host. To facilitate this the nematode has a much wider host range than usual.

served instead of caribou. As the meat was eaten dipped in seal oil the difference in taste was masked and, although the meat had been stored in the ice store for several months, the larvae survived and all the people who ate the dish were infected. More recently cases were reported from Idaho where salted and smoked cougar meat was eaten and resulted in ten of the 14 people who ate the meat experiencing symptoms of infection.

It is often stated that it is not in a parasite's interests to kill or be responsible for the premature death of its host as this will destroy its own habitat. As we shall see, in many indirect life-cycles damage to intermediate hosts may be advantageous to the parasite. *Trichinella* is one parasite where damage to and death of the definitive host acts in the parasite's favour.

Indirect transmission

So far our discussion has concentrated on parasites with a direct life-cycle. Parasites with two or more hosts have even greater need for efficient host location.

The large cestodes of man, *Taenia solium* and *T. saginata*, grow several metres long and release 8–9 proglottides a day containing up to 750,000 eggs. In the case of *T. solium* the proglottides are immotile, and the pig intermediate hosts are actively **coprophilic**. Cattle however do not feed close to their faeces and *T. saginata* segments are motile and can migrate onto surrounding pasture where they are more likely to be eaten. Once ingested the eggs hatch in the intestine and the oncospheres penetrate the gut wall. They are carried in the circulatory system to the somatic muscles where they grow and encyst as cysticerci (Fig. 3.9). Each cysticercus consists of a single scolex invaginated in a fluid-filled bladder. The cystic stages were known and named long before the whole life-cycles had been elucidated and they thus have names that do not necessarily bear any resemblance to the adult's name. The cysticercus of *T. solium* is known as *Cysticercus cellulosae* and that of *T. saginata* as *C. bovis*. The cycle is completed when the cyst is ingested. As much pork and beef production is for human consumption it is easy to see how transmission occurs. Thorough cooking and efficient freezing of the meat kills the cysts and effective meat inspection should remove infected carcasses from the human food chain. In many parts of the world none of these measures is in place and tapeworm infections are common. Even in Britain there has been an increase in *T. saginata* cases recently (*T. solium* which is a more serious problem as we shall see later, is not endemic). The reasons are not fully understood and the epidemiology is complicated (Fig. 3.10). The initial sources of infection are perhaps the most problematical but the increase in foreign travel and consequent increase in chances of becoming infected abroad are often blamed. Maintenance in cattle depends on human sewage reaching pasture. Direct contamination from cattle handlers and farm workers is one possibility but proglottides can survive in sewage and eggs may pass into sewage sludge which is regularly spread as a fertiliser. The pattern of infection of cattle does not accord with more or less even distribution of eggs, however, as only a few cows in a herd are infected. One possible explanation for this distribution is that whole proglottides get onto the pasture, but they should be broken down during sewage processing. Herring gulls feed widely and regularly around sewage farms and the motile proglottides may look like a tasty morsel to them. While the proglottides break down in the bird's gut the eggs can survive to be deposited onto the pasture in the faeces in discontinuous and relatively concentrated aggregations. The levels of infection in cattle are relatively low and may be missed at meat inspection. There is little risk if the meat is well cooked but steak tartare and rare steak may be a greater hazard. This makes infection an affliction of the affluent as they are the only ones who can afford the raw materials. This, of course, is not the situation in most of the world where it is

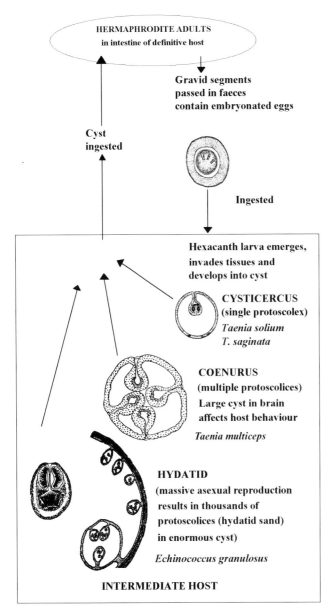

Fig. 3.9 Patterns of transmission of cyclophyllidean cestodes. The large cestodes *Taenia solium* and *T. saginata* produce hollow fluid filled cysticerci containing a single invaginated scolex. *T. multiceps* and *Echinococcus granulosus* produce larger cysts with multiple invaginated scolices. The size and site of the cyst may affect the behaviour of the host and make it more liable to predation.

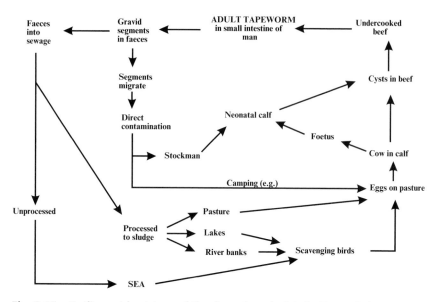

Fig. 3.10 Outline epidemiology of *Taenia saginata* in Britain. Transmission to man only occurs through eating infected beef, and cattle only become infected by ingesting eggs from human faeces. Pasture contamination may be direct from an infected person or from sewage. Scavenging birds may provide a route from sewage treatment plants to pasture.

the poor who cannot afford the cost of fuel to cook their meat who are most at risk.

Cysticerci are relatively small and each contains a single scolex. Unless infection is massive there is not usually any outward sign of infection in the intermediate host. This is not the case for all cestodes. *Taenia multiceps* grows to a 60 cm adult in the intestine of dogs. The basic life-cycle is similar to that of *T. saginata* with sheep as the intermediate host. In the sheep the oncospheres go to the central nervous system rather than the muscles and develop there. The cyst that forms is called *Coenurus cerebralis* and is considerably bigger than a cysticercus. The coenurus still consists of a fluid-filled bladder but instead of a single invaginated scolex forming inside it several hundred may occur. The intermediate host provides an opportunity for asexual reproduction to occur. Large cysts, up to 15 cm in diameter (see Fig. 7.3), developing in the brain affect the behaviour of the host. The exact effect depends on the precise site of infection but generally disorientation and often blindness and deafness occur. An infected sheep is much more likely than its flock mates to be preyed upon, which, from the parasite's point of view, is exactly what is required.

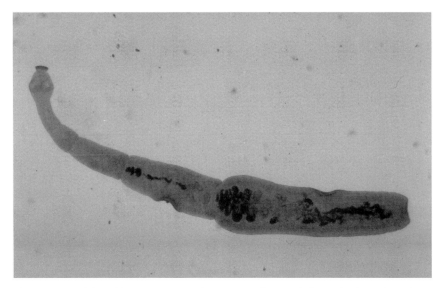

Fig. 3.11 Adult *Echinococcus granulosus*, a small cestode from the intestine of dogs. The adult usually consists of only 3–4 segments, one immature, one mature and one gravid. To make up for lack of adult reproductive potential the hydatid cyst that forms in the intermediate host may be massive and contain millions of protoscolices.

Another dog cestode, *Echinococcus granulosus*, has taken the idea of decreasing the reliance on sexual reproduction and increasing that of asexual reproduction in the intermediate host to its limit. The adult worms are only about 5 mm long and consist of a scolex and three proglottides, one developing, one mature and one gravid (Fig. 3.11). It takes about two weeks for a mature proglottid to become gravid, when it contains about 1,500 eggs, so the daily egg production is only about 100 compared with 750,000 for *T. saginata*. The diminutive *Echinococcus* makes up for its lack of adult stature by producing a massive cyst which has enormous reproductive potential. **Hydatid cysts** consist of a thick outer wall largely of host origin which surrounds a parasite-derived layer and a **germinal layer** from which small **brood capsules** bud off. Inside the brood capsules individual invaginated scolices develop. Because of the gritty nature of the cyst contents when rubbed between the fingers these **protoscolices** are sometimes called **hydatid sand**. Cysts containing 15 litres of fluid and millions of individual protoscolices have been reported. The cysts may form in a number of different organs but the liver and lungs are often infected. An intermediate host carrying several litres of cyst in its lungs is unlikely to be as capable of avoiding predation as uninfected animals and

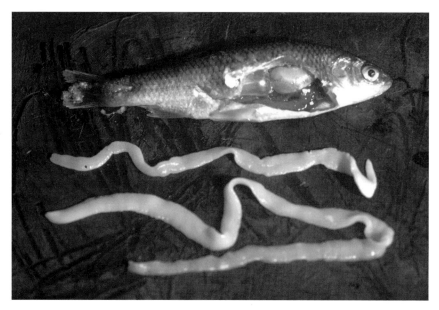

Fig. 3.12 Plerocercoids of *Ligula intestinalis* and the roach from which they were removed. The amount of parasite tissue may be as much as two-thirds of that of its host and cause disturbances in swimming behaviour which together with increased mortality in infected fish makes them more liable to be eaten by piscivorous birds. Photograph by Professor J. D. Smyth, reproduced with permission.

thus the chance of transmission is increased. The intermediate host can be one of a number of mammal species, including man, and, in different parts of the world, different epidemiological patterns occur depending on the local wild carnivore and herbivore species. Thus, in Australia it passes between dingoes and kangaroos, in North America between wolves and moose and elk, in Africa between wild dogs and goats and camels and in Britain between foxes and horses and sheep. In each place domestic dogs may replace the local wild dog and man may be a peripheral intermediate host. Because humans are not usually eaten by their dogs they do not normally contribute to the transmission cycle, but there are some African tribes that have such close associations with their dogs and such difficult problems burying their dead that many human hydatids get back into the dogs and man is a true intermediate host.

While it is the size and site of coenurus and hydatid cysts that makes the intermediate hosts of *T. multiceps* and *Echinococcus* more susceptible to predation, other parasites increase the conspicuousness of their hosts to achieve the same ends. The plerocercoid stages of the pseudophyllidean cestodes *Ligula* and *Schistocephalus* occur in the body cavity of fresh water fish, commonly roach (*Rutilus*) and sticklebacks (*Gasterosteus*) respectively (Fig. 3.12). The adults are found in the intestine of piscivorous birds. Compared with the mass of the fish plerocercoid tissue may be massive. The presence of this amount of parasite tissue affects swimming behaviour. Most fish are dark on the dorsal surface and light below. When swimming normally this colouring provides camouflage from above against the dark substratum and from below against the light surface. Infected fish find it difficult to swim normally and tend to wobble in the water. This destroys the camouflage effect as light is reflected off the lighter ventral surface and makes them more visible and liable to predation. Infection also puts extra stress on the hosts so that they are less able to survive starvation, and again their increased mortality makes them more likely to be eaten.

Infection of crustacean intermediate hosts with the cystacanths of acanthocephalans may also affect both conspicuousness and behaviour. The cystacanths themselves may be brightly coloured, often red, yellow and orange, and may be visible through the cuticle of the host. Other species alter the pigmentation of the host in a more fundamental way, making them either lighter, (*Acanthocephalus dirus* in the isopod *Lirceus lineatus*) or darker (*A. anguillae* in *Asellus aquaticus*) than uninfected animals. In feeding experiments up to 2.5 times more light-coloured infected isopods than normal-coloured ones were eaten by ducks.

Normally *Gammarus lacustris* are **photophobic**. They spend much of their time swimming on the underside and in the shade of aquatic vegetation and if disturbed their immediate response is to swim rapidly to the bottom and disappear into the mud. When infected with cystacanths of *Polymorphus paradoxus* they become much more **photophilic**. Upon disturbance they cling to vegetation and if there is no vegetation present they swim close to the surface of the water where they are much more vulnerable to predation by surface feeding duck definitive hosts.

Parasitic crustacea may also affect the behaviour of their hosts. Females of the parasitic copepod *Lernaeocera branchialis* (Fig. 2.14) lay eggs which give rise to nauplius larvae that have about a day to find and attach to the gills of a flounder (*Platichthys flesus*). On the flounder they produce an attachment thread which penetrates the skin of the fish and are known as **chalimus** larvae. After four moults the males become mature and transfer sperm to the females. Both sexes develop swimming **setae**, detach from the flounder and swim freely. The males die after mating but the females search for a member of the cod family. Once one is located they enter the gill chamber, attach near the fourth gill arch and begin to grow. The head penetrates the tissues of the host and produces ramifying branches that act as anchors while the body elongates, bends characteristically and bears little resemblance to most crustaceans. Infected cod spend a greater amount of their time in shallow water where flounders are found, thus increasing the chance of their parasite's eggs reaching their first host.

The first stage in the life-cycle of most digeneans is aquatic (Fig. 3.13). The eggs hatch in water releasing free swimming miracidia which invade the tissues of a mollusc. Bivalves filter miracidia from the water they pass through their gills but invasion of gastropods depends on more active host-seeking behaviour. Most miracidia show negative geotaxis and positive phototaxis when first hatched and initially show little response to the presence of snails. This behaviour is consistent with the miracidia being dispersive forms but it may bring them into the general environment of surface feeding snails and before long they begin to show an appreciation of snail presence and change their behaviour when in the presence of snail secretory products. The exact nature of the attractive stimuli is not known despite considerable investigation but chemicals likely to be components of mucus are generally suggested. Whatever the stimulus miracidia may be very efficient host locators. Miracidia of *Schistosoma mansoni* which are only about 500 μm long have been shown to be able to find snails at 9 m distance in still water and 100 m in gently flowing water. This can have serious consequences for control programmes as it means that a very small population of either parasite or intermediate host may be enough to maintain infection.

Following two phases of asexual reproduction the parasites leave the mollusc as free swimming cercariae. Some, such as schistosome cercariae directly penetrate intact skin, but most encyst somewhere in the food chain of their definitive host. For the liver fluke *Fasciola hepatica* this encystment occurs on vegetation and the encysted metacercariae await ingestion by a grazing animal. Although the miracidia and cercariae are essentially aquatic the digenean life-cycle allows plenty of scope for variation and this is well demonstrated by the other liver fluke found in the United Kingdom, *Dicrocoelium dendriticum*. *Dicrocoelium* is much smaller than the more common and widely distributed *F. hepatica* and in Britain is restricted to western Scotland and the Western Isles, but is widely distributed in Europe and the USA. *Dicrocoelium* eggs are embryonated when passed in the faeces and hatch on ingestion by a terrestrial snail, *Zebrina detrita* among others, which feed on animal droppings.

The miracidia emerge in the snail's gut, penetrate into the digestive gland and, following two sporocyst stages, produce cercariae which are expelled from the snail in 'slime balls' consisting of mucus produced partly by the cercariae themselves and partly by the snails in response to the irritation caused by the migrating cercariae. The slime balls are sticky and adhere to the herbage

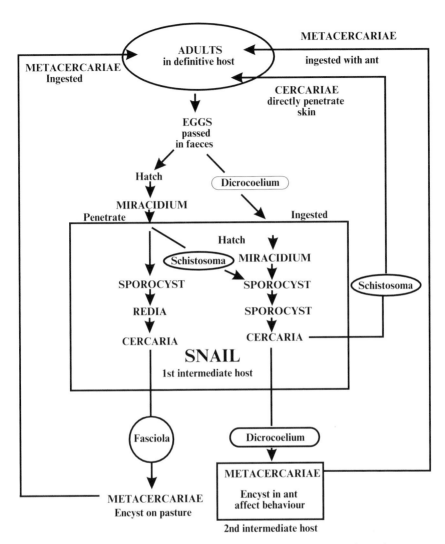

Fig. 3.13 Transmission patterns of some digenean parasites. *Fasciola* and *Schistosoma* have life-cycles that rely on water for infection of the snail intermediate host and for the cercariae to leave the snail to encyst or penetrate the skin. *Dicrocoelium* has adapted a terrestrial life-cycle with ingestion of embryonated eggs by the snail and transfer to the second intermediate host by ingestion of cercariae in slime balls. The metacercariae in the ant secondary host may affect its behaviour and make it more susceptible to ingestion by a sheep.

where they are something of a delicacy for wood ants, *Formica fusca*. In the ant the cercariae encyst in the tissues as metacercariae ready for transmission to another host. One or more of the metacercariae encyst in the suboesophageal ganglion of the ant and affect its behaviour. Instead of returning to the nest at the end of the day infected ants remain with their mouthparts locked towards the top of the vegetation where they are more likely to be eaten by a grazing animal. *Dicrocoelium* has thus taken an essentially aquatic life-cycle and adapted it to be entirely terrestrial.

At first sight it might be thought that there is little opportunity for parasites that rely on blood feeding vector insects for transmission to show adaptations for transmission. All they should need to do is hang around and wait to be picked up. This may be true for many protozoans but is not the case for all parasites. There is evidence that *Onchocerca* microfilariae are attracted to the salivary secretions that are injected into the wound made by their *Simulium* vectors. Fed black flies contain more larvae than would have been expected from the population in the skin. Microfilariae of *Wuchereria bancrofti* display adaptations not to the presence of its night-biting mosquito host but to its behaviour. For most of the day the microfilariae are found in the capillaries of the lung with few in the peripheral blood, but at night they emerge and are found in the general circulation. Strains of the parasite that are transmitted by day-biting mosquitoes show modified behaviour with more or less reversed periodicity. The most important factor controlling the parasite's behaviour appears to be the oxygen gradient in the lung capillaries. During the day the gradient is steep, the microfilariae are active and do not pass easily through the finer lung capillaries. At night the gradient is flatter, the microfilariae become less active and are carried passively through the capillaries into the circulation where they are available for feeding mosquitoes (Fig. 3.14).

Arthropod adaptations

And how do vectors locate their hosts? Most insects have highly developed sense organs and sight, olfaction, temperature and vibration act as attractive stimuli or indicators of host presence for different species. Temperature and carbon dioxide are general long-distance cues for many species although most exhibit a hierarchy of host preference and more specific cues are recognised; tsetse flies (*Glossina morsitans*), for example, are attracted to acetone in the exhaled breath of cows.

Three host hard ticks spend the majority of their lives on the pasture with

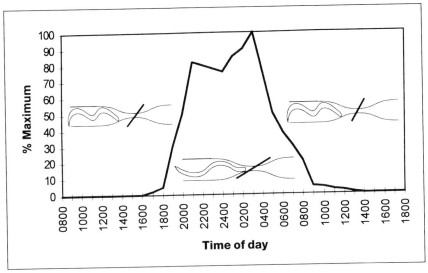

Fig. 3.14 Periodicity of *Wuchereria bancrofti*. During the day, when the oxygen gradient in the finest capillaries in the lungs is steep, the microfilariae are active and do not pass. At night the gradient is flatter, and the microfilariae are less active and are swept through into the systemic circulation where they are available for night biting mosquito vectors.

only three comparatively short periods feeding. After hatching or moulting the time off the host is spent moving up and down grass stems in response to changes in humidity. During periods of high humidity they are found near the top of the stems extending their first pair of legs in a questing posture hoping to encounter a passing host. When the humidity decreases they retreat down the stem to conserve water, repeating the process until they locate a host or die in the attempt. Actual contact is a very hit or miss affair.

Fleas show more sophisticated responses to the presence of hosts. In temperate regions bird fleas often spend the winter in their cocoons, emerging when the weather warms up in the spring. Initially they tend to be negatively phototactic and remain in the nest mating. After a few days their behaviour changes and they become negatively geotactic and positively phototactic, climbing out of the nest and onto vegetation. The presence of a potential host is signalled by a sudden drop in light intensity, when the fleas launch themselves with their second pair of legs extended over their backs like grappling hooks in the hope of attaching. If they fail they land on the ground and begin the long climb back up to repeat the process. Mammal fleas are especially sensitive to smell although warmth and carbon dioxide are also attractive. Vibration and air currents may also trigger a leaping response. It is often the vibrations of entry into an empty property that stimulate the fleas to leave their cocoons to greet the new owners. The process may appear somewhat haphazard but large numbers achieve their target. In one experiment nearly half of 270 marked rabbit fleas released in a 2000 square metre area were recovered from three rabbits within a few days. The overall effectiveness of parasite transmission is illustrated by the enormous number of hosts that carry at least one, and often many parasitic companions.

4

Getting settled

The adaptations discussed in the previous chapter encourage contact between parasite transmission stages and potential hosts, but the problems do not end there. The host may not be suitable, and even if it is it may present barriers that must be overcome before establishment can be achieved.

There are essentially two portals of entry into a host, through the mouth or through the skin. Many parasites use the **oral route of infection** to enter both definitive and intermediate hosts. Direct **percutaneous infection** is less common but two important groups of helminth parasites, the schistosomes and the hookworms, make use of it, as do the blood feeding ectoparasitic arthropods. These latter provide an entry route for vector transmitted protozoans and helminths.

Oral infection

The alimentary canal (Fig. 4.1) offers a wealth of habitats for parasites, although some regions are more hospitable than others. The mouth is the obvious portal of entry and a parasite that can get its infective stages into the food chain of its host has a good chance of achieving transmission.

The oral cavity and oesophagus tend to be sparsely populated by parasites, although some specialists have adapted to them. Both regions have an essentially protective epithelium which, in many species although not man, is keratinised to protect the surface from damage by poorly masticated coarse and rough food. Food spends little time in them but, lubricated with saliva, passes rapidly to the stomach. Among the parasites that are found in the upper

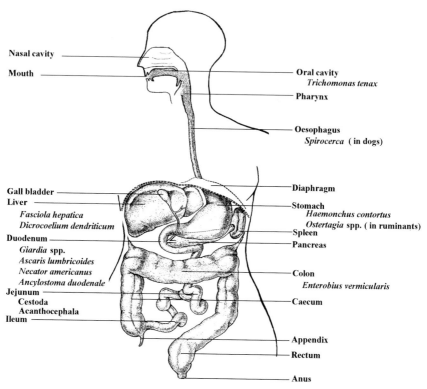

Fig. 4.1 Human (mammalian) alimentary canal with a few of the parasites associated with the various regions.

alimentary canal are *Trichomonas tenax* (Fig. 2.2) which lives in the oral cavity and is transmitted directly, and *Spirocerca lupi*, a nematode that lives beneath the epithelium of the oesophagus of dogs where it may result in the formation of malignant tumours. Birds have a crop, an expanded sac formed at the base of the oesophagus, that provides a relatively quiescent reservoir in which *Trichomonas gallinae* can flourish.

Relatively few parasites establish in the stomach, where the combination of high proteolytic activity, low pH and rapid cell turnover make it an inhospitable environment. The stomach may also act as an impassable barrier to any invader not suitably protected from its chemically active contents. The stomach of ruminants consists of four chambers, the rumen, the reticulum, the omasum and the abomasum. Only in the last of these does proteolytic digestion occur. The first three, especially the rumen, contain the bacteria and protozoa that provide the cellulase that initiates cellulose digestion.

Parasites such as the nematodes *Haemonchus contortus* and *Ostertagia*

(= Teladorsagia) circumcincta that can survive in the abomasum of sheep are pathogenic as they interfere with gastric function and thus upset the digestive process at its outset. This reduces feed conversion rates and weight gains and has economic consequences which run into millions of pounds per annum for animal producers.

The intestine is the happy hunting ground for the enthusiastic parasitologist. From duodenum to rectum parasites abound. Virtually all cestodes and acanthocephalans are found in the intestine together with innumerable protozoans, digeneans and nematodes. The intestines provide warmth, shelter and, assuming the host is reasonably nourished, a plentiful supply of already partially processed food. There may be some instability in the substrate due to peristaltic contraction but this is more or less predictable, rarely very violent and most parasites have means of remaining in place (see Chapter 5).

The different regions of the gut are not homogeneous and, especially through the length of the intestine, physico–chemical gradients occur which vary throughout the day and with the passage of food (Fig. 4.2). The subtle differences that these present to parasites are poorly understood but there is little doubt that parasites recognise them. The European tortoise (*Testudo graeco*) acts as host for up to eight species of oxyurid nematodes (*Tachygonetria* spp.) which, on superficial examination, appear to live in glorious confusion throughout the colon. Critical analysis of the three-dimensional distribution of the individual species demonstrated that even in this limited environment each species had its preferred niche (Fig. 4.3). Some species are essentially lumen dwellers while others frequent the mucosal surface, some feed on bacteria while others ingest gut contents and, where two species appear to have requirements that overlap appreciably, temporal differences ensure that the two species are not present at the same time.

Parasite transmission stages are generally relatively long-lived, with a low metabolic rate to conserve restricted energy reserves. How do they recognise that it is time for them to resume development?

In mammals and birds temperature is an obvious cue and is certainly important, but it is so general that if it was the sole factor development might well start too early and the young parasite be destroyed in the stomach before it has a chance to reach its favoured site in the intestine. That said, the metacercariae of some digenean parasites of the mouth and oesophagus of birds excyst simply in response to a temperature increase above 35°C.

Physico–chemical cues provide the other major stimuli and these differ for different groups of parasites and different species.

Hatching and exsheathment of those nematodes that have been examined is controlled largely by carbon dioxide at an appropriate pH at 37°C. Changing

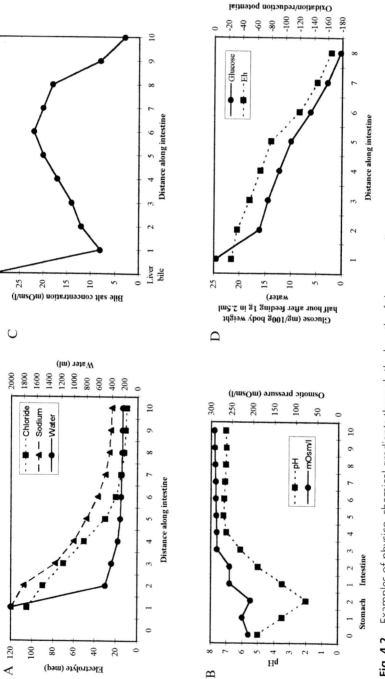

Fig. 4.2 Examples of physico-chemical gradients through the length of the mammalian gastro-intestinal tract. A, water and electrolytes. B, pH and osmotic pressure. C, bile. D. glucose and oxidation/reduction potential. Data from: Mettrick, D. F. & Podesta, R. B. (1974) *Advances in Parasitology*, **12**, 183–278; Read, C. P. (1971) in *Ecology and Physiology of Parasites*, Hilger.

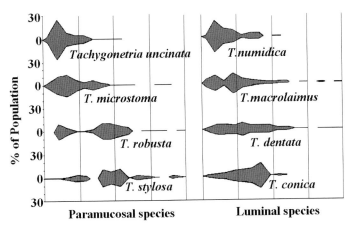

Fig. 4.3 Distribution of *Tachygonetria* species in the intestine of the Greek tortoise. Of the eight species illustrated four are found predominantly in the gut lumen while the other four are concentrated close to the mucosa. In each region the different species have different longitudinal distributions and no two species occupy exactly the same niche. Redrawn from Schad, G.A. (1963). *Nature*, **198**, 404–406.

other factors such as the oxidation/reduction potential may enhance the effect but CO_2 appears to be the essential component. In most animals the first site that the parasite meets that fulfils these requirements is the duodenum, although the first stomach, the rumen, of ruminants may supply similar conditions for some of their parasites. The correct stimulus is thought to trigger the production or release of hatching or exsheathing enzymes which, once initiated, continue to work in the absence of further stimulation. Pre-treatment with pepsin or other host proteolytic enzymes does not enhance the process. For example *Ascaris* eggs are unembryonated when passed in the faeces of their host. They require up to three weeks at 30°C to embryonate and develop to the second stage larva before they will hatch (Fig. 4.4A). *In vitro*, in the presence of CO_2 at a temperature of 37°C eggs hatched preferentially at a slightly alkaline pH (Fig. 4.4B). In response to these stimuli the larvae release a lipase, a chitinase that can be measured by determining the level of the chitin breakdown product N-acetyl glucosamine in the incubation fluid (Fig. 4.4C), and probably a protease. These result in the dissolution of a small area of the egg shell from which the larvae, still enclosed in the inner lipid layer of the egg shell, protrude. Eventually they rupture the lipid layer and escape from the shell (Fig. 4.4D). *Ascaris* eggs are notorious for their impermeability, which is largely dependent on the chemical structure and integrity of the lipid layer. For the enzymes released by the larvae to reach the

A.

Days at 30°C	% Hatch
17	1
18	2
19	78
20	82

B.

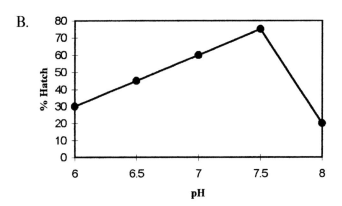

C.

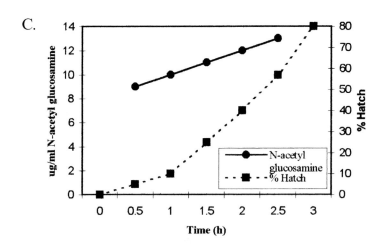

D.

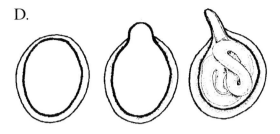

E.

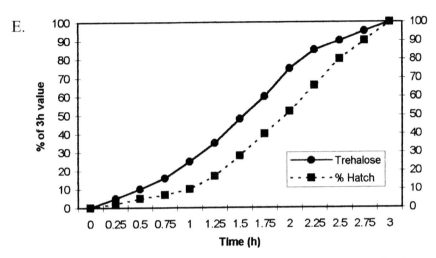

Fig. 4.4 Physiology of hatching of *Ascaris lumbricoides*. A, the eggs only hatch after complete embryonation and development of the second stage larvae within the egg shell. Once triggered by the presence of CO_2 hatching is influenced by pH, B, and can be monitored by measuring the chitin breakdown products in the bathing fluid, C. Initially the permeability of the inner egg membrane increases allowing enzymes produced by the larva access to the inner shell surface, monitored by the level of trehalose in the supernatant, E, and eventual digestion of a hole through which the larva emerges, D. Data from: Rogers, W. P. (1974) *Nature*, **181**, 1410–11; and (1960) *Proc. Royal Soc. B*, **152**, 367–86; Fairburn, D. (1960) in *Host Influence on Parasite Physiology*, Rutgers University Press.

inner layer of the egg shell the permeability of the lipid layer must first be altered. This change can be followed by monitoring the concentration of the sugar trehalose in the medium bathing the eggs. In unstimulated eggs the larvae are bathed in fluid containing a relatively high concentration of trehalose. After stimulation the trehalose begins to leak out of the eggs followed about half an hour later by hatching (Fig. 4.4E).

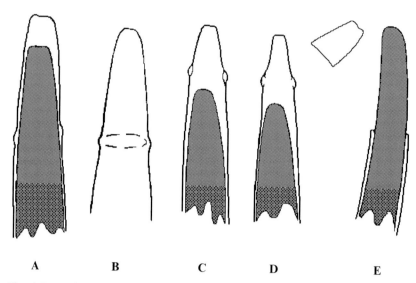

A B C D E

Fig. 4.5 Exsheathment of nematode larvae. The stimulus for exsheathment varies from species to species but once initiated secretions from the excretory pore cause the formation of a refractile ring, A and B, around the anterior of the sheath. At this point the sheath splits into two layers, C, the inner layer ruptures, D, and the anterior is lost as a cap. Redrawn form Lapage, G. (1935). *Parasitology*, **27**, 186–206.

All nematodes moult between each developmental stage. The old cuticle first separates from the hypodermis which secretes a new cuticle beneath it. The old cuticle is normally then dissolved, partially resorbed or breaks down. In some species the final stage in this moulting process is delayed and the old cuticle is retained as a sheath around the worm where it provides additional protection from desiccation.

Ensheathment is most commonly found among the infective third stage larvae of bursate nematodes and the first phase in the infection process for these species is exsheathment, the loss of the sheath. Following stimulation by CO_2 at an appropriate pH the sheath swells near the anterior end (Fig. 4.5A&B), and the wall separates into two layers (Fig. 4.5C), the inner of which is digested leaving a weakened ring (Fig. 4.5D). The activity of the enclosed larva causes the sheath to rupture around the weakened ring releasing the anterior as a cap and liberating the larva. The source and nature of the exsheathing fluid has been the cause of some controversy over the years. The general consensus at present seems to be that exsheathing fluid originates in the excretory cell, is released through the excretory pore and is chemically characterised as a pseudocollagenase.

The hatching of taeniid cestode eggs and the excystment of both digenean and cestode cysts appears to be a two stage process. The first stage is stimulated by temperature and relies on host enzymes to remove the outer cyst membranes or to degrade the cementing material between the component blocks of the **embryophore**. The second phase is stimulated by bile which again limits the excystment/hatching site to below the entry of the bile duct in the duodenum. Bile changes the permeability of the remaining membranes and initiates active movements that lead to final emergence.

The exact role of the individual components of bile has not been fully resolved but there is evidence that in some cases individual bile acids may contribute to host specificity. For example, the small tapeworm, *Echinococcus granulosus*, normally develops in the intestine of the dog and emergence from the hydatid cyst is stimulated by dog bile which contains 177 mg/100 ml deoxycholic acid. Rabbit bile contains 3600 mg/100 ml and is inimical to juvenile *Echinococcus*, resulting in their dissolution and destruction.

Those helminths that hatch in the intestine of invertebrate intermediate hosts have less exacting requirements. For example, acanthors of the acanthocephalan *Moniliformis dubius*, which normally hatch in the gut of cockroaches, can be induced to emerge from their eggs *in vitro* by 0.25 M sodium chloride solution; a pH above 7.5 and the presence of CO_2 enhance the effect. Brief exposure to the electrolyte solution is sufficient to stimulate vigorous activity in the acanthor and the action of the rostellar blades combined with the release of a chitinase break down the shell and allow the acanthor to emerge.

Hymenolepid cestode eggs which hatch in a number of arthropod hosts depend on mechanical damage to the outer egg membranes from the insect's mouthparts to cause initial changes before changes in ionic composition stimulate activity of the hexacanth and emergence from the remaining membranes.

Coccidian oocysts also have a two stage excystation process. The first, initiated by CO_2, causes changes in the permeability of the oocyst wall and allows trypsin and bile, which constitute the stimulants for the second phase, to reach the sporozoites. These are activated to escape from the weakened oocyst. The site of emergence of sporozoites from the oocyst governs the initial site of establishment in the host gut (Fig. 4.6). Species which excyst rapidly establish higher up the gut than those that excyst more slowly. The situation is, however, more complicated as oocysts experimentally introduced through an unusual route, e.g. intravenous injection, establish patent infections in the normal gut site. *In vitro*, *Eimeria* species are able to develop in a wide range of cultured cells but *in vivo* are largely restricted to gut epithelial

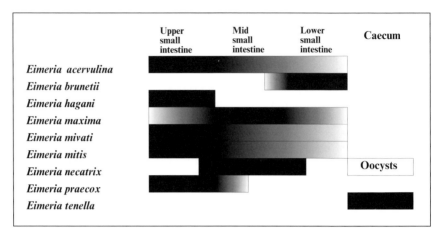

Fig. 4.6 Distribution of *Eimeria* species in the intestine of the chicken.

cells and the final site is determined by at present poorly understood host factors.

Percutaneous infection

Infection through the skin may be directly through the intact tissue or through a lesion usually made by the mouthparts of an arthropod vector.

Schistosomes and hookworms penetrate mammalian skin directly. Approach to the skin is up a temperature gradient and there is evidence that skin lipids stimulate penetration behaviour, but only at a very short range and possibly not until contact has been made. Schistosome cercariae are equipped with two sets of glands which open into the oral sucker (Fig. 4.7A). Once the skin has been reached cuticular extensions around the oral sucker abrade the surface and secretion from the preacetabular glands begins. These secretions are alkaline and sticky and act to hold the cercaria to the skin and also to soften the keratinised cells of the **stratum corneum**. Entry into the skin is through a desquammation or wrinkle (Fig. 4.7B) and may be, but does not have to be, through a hair follicle. Once established in the skin the postacetabular glands begin to release their secretions (Fig. 4.7 C&D). These are proteolytic and digest the cells of the **epidermis** (Fig. 4.7E) to allow the cercaria, now transforming into a **schistosomulum**, to enter the **dermis** (Fig. 4.7F) and thence to the circulatory system.

Some hookworms, e.g. *Necator americanus* (Fig. 4.8A) also produce proteolytic secretions from their **oesophageal glands** to ease entry into the skin

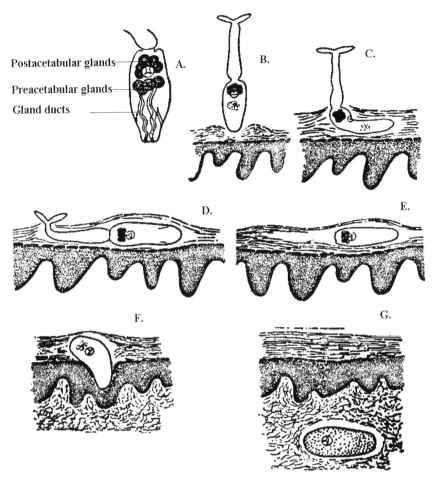

Postacetabular glands

Preacetabular glands

Gland ducts

Fig. 4.7 Process of invasion of the skin by *Schistosoma mansoni* cercariae. Approach to the skin is up a temperature gradient. Once the surface is reached sticky preacetabular gland secretions help to hold the cercaria in place. Exploration of the surface leads to entry at a defect in the outer skin surface and subsequent release of proteolytic enzymes from the postacetabular glands to digest a passage through the epidermis and into the dermis. Adapted from Gordon, R. M. & Griffiths, R.B. (1952). *Annals of Tropical Medicine and Parasitology*, **45,** 227–243.

of their hosts and their passage is marked by a trail of damaged cells. Other skin penetrating nematodes, e.g. *Ancylostoma tubaeforme* (Fig. 4.8B) and *Nippostrongylus brasiliensis*, do not have such enzymatic armaments and appear to be able to penetrate mechanically, possibly with the aid of ions carried in with them to upset the electrostatic forces that hold the epidermal cells

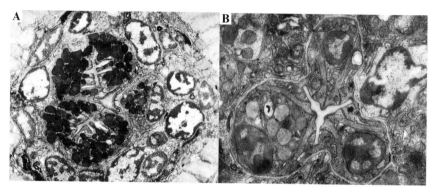

Fig. 4.8 Electron micrographs of the oesophageal glands of third stage infective larvae of A, *Necator americanus* and B, *Ancylostoma tubaeforme*. The large and more active glands in *Necator* lend credence to other studies which suggest that *Necator* penetrates the skin enzymatically while *A. tubaeforme* does so mechanically. Photographs by Smith, J. M. (1976). *International Journal for Parasitology*, **6**, 9–13.

together. Once these species have passed through the skin the cells recombine and there is no visible lesion.

By their very nature blood feeding ectoparasitic arthropods need to breach the integument of their hosts to obtain their blood meals. To prevent their hosts' blood clotting many inject an anticoagulant produced in their salivary glands into the wound. The passage of saliva provides an ideal vehicle for parasites that can invade the salivary glands to bypass the barrier presented by the host's skin and obtain direct entry into the blood or tissues.

Most ectoparasites are host specific or at least show host preferences, so that if a parasite has been picked up from one definitive host there is a good chance that, if the vector lives long enough, it will feed on another suitable animal to pass the parasite to its next definitive host. Vectors respond to the **host stream**, a plume of host-conditioned air that emanates from the host and extends downwind. Both group factors, for example CO_2 which is universal to all vertebrates, and host specific factors (odours and other chemical products) contribute to host selection. While relying on vectors for transmission removes some of the direct problems of locating a new host it is not without its problems for both parasite and vector. Not all species of a given genus are equally suitable as vectors, for example only some 70 of the 400 species of *Anopheles* mosquitoes are malaria vectors. Some may take only a single blood meal and thus act as a dead end while others may not live long enough for the parasite to complete its developmental stages, and, of course, some parasites will be taken up during the last feed that the vector takes and die with it. The parasite may damage the vector to the detriment of both. Fifteen *Wuchereria* microfilariae per 20 cubic millimetres of blood, the normal volume taken for diagnostic purposes, is ideal for transmission of the nematode through its mosquito vector; 100 microfilariae per 20 cubic millimetres is fatal to the mosquito. Up to 600 microfilariae per 20 cubic millimetres has been recorded in the peripheral blood of patients at the peak of infection.

Protozoan parasites are small enough to pass down the proboscis of their vectors and are thus injected into their new hosts with the saliva. The infective third stage larvae of filarial nematodes are too big to pass through the lumen of the proboscis. They break out of the mouthparts and are released on to the skin surface where they invade either the wound made by the vector or another lesion.

Once in the blood stream of their new definitive host one might expect that blood parasites would be satisfied and settle down to a new life. Many do so but the malaria parasite *Plasmodium* is first carried in the circulatory system to the liver where it invades the liver cells and undergoes asexual reproduction. Exactly how the parasite realises that it has reached the liver is still not known but it is thought that it has specific molecular receptors on the cell wall that chemically recognise complementary molecules on the liver cells. This is similar in principle but differs in detail from the way in which the merozoites that are later released from the liver invade red blood cells. Merozoites (Fig. 4.9A) have a thick cell coat covered with fine filaments with T- or Y-shaped endings. These adhere to specific receptors on the red cell surface. The merozoite can attach at any point on its surface and contact with an erythrocyte is probably random and fortuitous. Once attached it orientates itself until the apical prominence comes into contact with the red cell membrane (Fig. 4.9B) when the release of products of the rhoptries and apical complex cause the cell wall to invaginate allowing the merozoite to enter the vacuole formed (Fig. 4.9C). The merozoite loses its cell coat in the process and the parasite comes to lie in a membrane bound parasitophorous vacuole. The initial recognition and attachment of the merozoite to the red cell is critical and highly specific. The surface membranes of the red cells of most human ethnic groups contain molecules known as the Duffy antigens. These are lacking in West African Negroes who are also resistant to infection with *Plasmodium vivax*. The two facts are not unrelated. Duffy negative erythrocytes lack the specific recognition receptors for *P. vivax* merozoites which are unable to adhere to or invade them. Despite considerable effort, and some encouraging leads, no such clear cut evidence exists for mechanisms blocking erythrocyte entry by other malaria species, although the search continues.

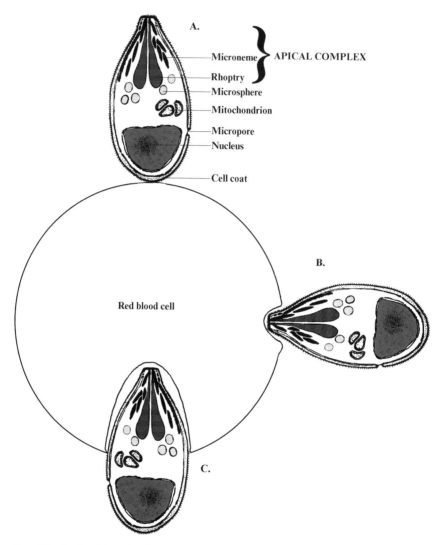

Fig. 4.9 Red cell invasion by merozoites of *Plasmodium*. The merozoite adheres to the red cell membrane by its cell coat, orientates itself, B, and causes an infolding of the cell membrane by secretions from the apical complex.

While more is known about malaria invasion of red blood cells because the system lends itself to experimental investigation, similar mechanisms probably apply to other sporozoans such as *Eimeria*. It is not too difficult to postulate that relatively small differences in receptor chemistry could account for the cell specificity seen in different species.

Table 4.1. *Site and exsheathment conditions for bursate nematodes of sheep*

Species	Exsheathment site	pH for exsheathment	Maturation site
Haemonchus contortus	Rumen	Neutral	Abomasum
Ostertagia circumcincta	Rumen	Neutral	Abomasum
Trichostrongylus colubriformis	Abomasum	Acid/pepsin	Small intestine
Nematodirus spp.	Abomasum	Acid/pepsin	Small intestine
Oesophagostomum columbianum	Small intestine	Neutral/alkaline	Large intestine

Migration within the host

The site of activation and emergence from their resting stages is often not the site of maturation for the adult parasite. Nematodes frequently exsheath at a site anterior to their final adult site (Table 4.1). Many species then spend part of their developmental time within their hosts' tissues, either in the gut wall or while undertaking complex migrations before attaining their preferred adult site. These histotrophic stages are often significant in causing pathological changes that result in clinical disease for the host.

The classic tracheal migration route, as exemplified by *Ascaris* and related ascarid nematodes (see Chapter 3) has been seen as almost universal for nematodes and has been accepted as the standard route for skin penetrating species as well as species that start their journey in the intestine. It has the advantage that it feels intuitively correct. The larvae are following lines of least resistance which allows the host to do much of the work of transportation and conserves the parasite's essential energy resources. Recently the widespread acceptance of this route has been questioned. Evidence for the route has been based on damage to the organs concerned in hosts that have been heavily experimentally infected, and also the finding of larvae in the organs on the proposed route sequentially after infection. It has been argued that this **sampling at autopsy** is not valid as it demonstrates the points on the route where the larvae are impeded in their passage and have possibly diverted from the normal route, and not the points they are passing through easily and rapidly. In any journey one spends a relatively longer time lost on winding country roads than on the motorway (the M25 excluded!). Critical numerical analysis of relatively light infections in rats of two skin penetrating nematodes, *Nippostrongylus brasiliensis* and *Strongyloides ratti*, have shown that the former does indeed use the tracheal route to reach the intestine but that *Strongyloides* larvae get there *via* the head. The precise route to the head has still to be resolved although migration of the larvae within the dermis has been suggested.

The tracheal route is by no means universal among nematodes that migrate to their adult sites. Cattle are infected with the lungworm *Dictyocaulus viviparus* when they ingest their larvae with herbage. The larvae are unusual as they do not feed but live on reserves obtained from the egg. They are also unusually sluggish, perhaps reflecting their need to conserve their energy, and unlike other infective nematode larvae, do not migrate away from the faecal mass in which they have developed. As cattle do not graze the pasture immediately around their droppings the larvae need to get away from them and they have adopted an ingenious way to preserve their energy and still achieve translocation from the faeces. They await the growth of a fungus, *Pilobolus*, which is very commonly found on cow pats, climb the sporangia and when these burst explosively the larvae are launched away from the faeces onto the pasture. Once ingested and in the intestine they penetrate the intestinal wall and enter the lymphatic system, passing through the mesenteric lymph nodes on a journey *via* the thoracic duct and the heart to the lungs.

Adult *Strongylus vulgaris* live in the large intestine of horses where they suck a plug of mucosa into their large buccal capsules and feed on mucosal debris and blood. The damage and blood loss may lead to unthriftiness and anaemia but is not life threatening. Experimentally, at least, the migrating larval stages may be more of a problem. Third stage infective larvae are ingested with herbage and invade the intestinal mucosa where they moult to the fourth stage. These larvae enter small arteries and migrate to the cranial mesenteric artery and its main branches. In the arteries they cause large **thrombi** to form around them; these may detach and cause fatal blockage of the coronary artery – an equine heart attack. Alternatively the arterial wall may be weakened and damaged resulting in **aneurysms**. As mentioned, these catastrophic consequences normally result from heavy experimental infections and natural exposure rarely causes such problems. In these natural infections the fourth stage larvae, after a period of several months in the mesenteric artery, return in the circulatory system to the submucosa of the caecum and colon and thence to the lumen and maturity.

Not all nematodes migrate or even invade their host's tissues before maturing. Experimental infections where *Trichinella spiralis* larvae were surgically introduced into different parts of the gut of mice demonstrated that the adults developed at the site of introduction and did not move to the anterior small intestine where they are normally found in simple oral infections. It was concluded that what is normally seen as the typical adult site just reflects the point at which the larvae are digested from the inoculum. Experimentally changing such factors as the age of the host, gut motility, size of inoculum and the size of the vehicle of infection all influenced the final adult site. It should be remembered that *T. spiralis* has an abnormally wide host range and may thus be less specific in its requirements; few species are so accommodating.

Stimulated by the presence of bile *Fasciola* metacercariae excyst in the duodenum. The juvenile flukes penetrate the gut wall using a combination of proteolytic enzymes and cell ingestion and enter the peritoneal cavity. They need to locate the liver in order to complete their development and a number of theories have been propounded to explain how they get there. The first assumes that they recognise the direction of the liver from chemical gradients emanating from it and are attracted up them. A second and more recent suggestion is that the gut and the liver both produce factors that affect the activity of the young flukes. The gut products are thought to increase activity and the liver products to decrease it. Close to the gut the flukes move rapidly and thus tend to move away, then, as they approach the liver, their activity decreases and they remain close to it, eventually making contact and ingesting their way through the liver capsule and into the bile ducts. A third idea is that, once in the closed peritoneal cavity, migration in any direction will eventually lead to the liver. Directionality and sophisticated sensory responses are not required for this latter hypothesis and, as appropriate sense organs and attractants have not been demonstrated *in vitro*, it is especially attractive.

Hymenolepis, and probably other cestode hexacanth larvae, initially establish about mid-way along the small intestine of their hosts. As the strobila begins to develop the scolex moves forwards and the posterior backwards so the midpoint of the worm remains in a more or less constant position in relation to the length of the intestine. Large cestodes and acanthocephalans may overlap a number of regions of the gut and the idea of strict site specificity may not be as appropriate for them as for many other parasites, a topic we will return to in the next chapter.

5

At home with the host

Once established inside its host the parasite is protected from the vagaries of the climate that torment free-living organisms. The host provides a dark, constantly warm (in birds and mammals), comparatively stable environment with an ample and more or less continuous supply of nutrients. This internal environment is not without its hazards: physically the intestinal tract is in continuous motion and can displace and eject any unwary or unprepared parasite, chemically the wide pH range and variety of proteolytic enzymes may be disastrous for unprotected organisms, and the host's immune response to the presence of the parasite may make its position untenable. The problems of finding a suitable host are prodigious and having found one the parasite wants to hang on to it and not let it go without a struggle.

At the end of the last chapter we considered parasite migration during establishment in its **predilection site** and the idea of fixed site specificity was questioned. Before discussing how parasites manage to exploit their internal environment we first need to return to site specificity as it is crucial to the later topics.

The word 'site' to represent the restricted locality in or on a host is deeply entrenched in parasitological literature. It is however in some ways misleading as it implies a fixed and unalterable region. For some parasites at least, experimental or natural changes in the host may alter the position of the parasite. For example, in the first week or so of a primary infection in normally fed rats, the nematode *Nippostrongylus brasiliensis* is found in a restricted area about 30% of the way along the 100 cm of the small intestine. If the host is fasted they are more evenly distributed throughout the length of the intestine (Fig. 5.1A). Careful analysis of the position of the worms in the gut shows that they are

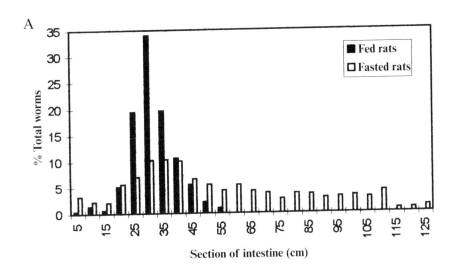

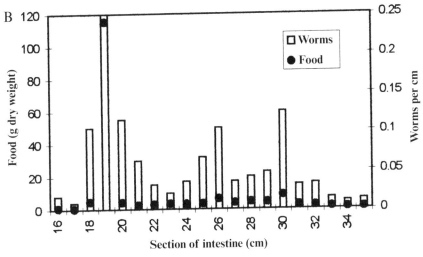

Fig. 5.1 Distribution of *Nippostrongylus brasiliensis* in the small intestine of fed and fasted rats and the relationship between the position of the worms and food in the gut. Redrawn from Croll, N.A. (1976). *International Journal for Parasitology*, **6**, 441–448.

highly correlated with the food as it passes down the gut (Fig. 5.1B). The worms feed on the gut contents which, at about the 30% point, have the correct consistency for ingestion. In fasted hosts the worms are associated with the mucosa living among the villi, no doubt seeking sustenance. *Nippostrongylus* infections in rats are generally relatively short lived and self limiting as the host mounts an immune response against their presence which eventually eliminates them (see Chapter 6). As a result of this during the third week of the infection the nematodes are found both further forward and further back than the 30% 'optimum site' as this becomes less ideal. As more examples like this have come to light there has been a move away from the idea of a single rigid 'site' for each parasite and it has been replaced by the more general concept of an ecological 'niche' to describe the four-dimensional space within the internal environment that the parasite occupies. That said, 'site' is likely to remain in use as a shorthand term amongst parasitologists for the foreseeable future.

As we have seen, the precise position of *Nippostrongylus* in a given individual host is dependent on the feeding cycle. In the laboratory, rats normally have unrestricted access to food and the worms are generally found in their normal 'site' but, in the wild, rats are mainly nocturnal feeders and their food supply may be more erratic. Parasites under these conditions may respond to changes in the host gut in different ways. The best-worked example is probably that of the cestode *Hymenolepis diminuta*. The developmental, **ontogenetic migration** of the scolex from its establishment site in the mid-intestine to the adult site in the anterior intestine was described in the last chapter. In addition to this *H. diminuta* undergoes a daily, **circadian migration** associated with the presence of food in the gut. When the stomach is full, usually in the morning after a night's feeding, the scolices are found within the first 15 cm of the **pyloric sphincter**. If food is withheld the stomach is empty by about midday at which time the scolices begin to migrate posteriorly to finish some 30 cm down the small intestine by mid-afternoon (Fig. 5.2). The terminal segments of the worm also move posteriorly and the whole worm responds to the physiological state of the host's intestine. If the feeding regime of the rats is experimentally reversed so that they have access to food by day only then the pattern of migration is partially but not completely reversed. This suggests that the relationship between feeding and migration is not a simple one.

The hypothesis is that *H. diminuta* attempts to maintain the majority of its strobila within the optimal zone for nutrient absorption. This may be achieved by both scolex migration and muscular activity which extends or shortens the length of the strobila, folding it and keeping its bulk in the optimal position. Cestodes are not highly motile organisms capable of vigorous muscular activity, however they do not lie passively in the gut but respond to changes in their environment. The hypothesis postulates sensory receptors along the length of the worm to detect the changes but these have not yet been unequivocally demonstrated. The posterior migration may just be

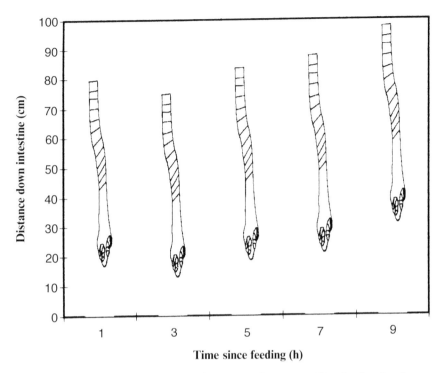

Fig. 5.2 Migration of *Hymenolepis diminuta* in the rat gut. Shortly after feeding the scolices of the worms are found in the first 20 cm of the intestine. As the stomach empties the scolices move posteriorly. At the same time the worm elongates and maintains contact with the food for the longest possible time. Data from Hopkins, C.A. (1970). *Parasitology*, **60**, 255-271.

passive, with the worms selecting an optimal zone when the gut is full and then drifting backwards with peristaltic movements, or they may select optimal positions throughout the feeding cycle so that the posterior migration is positive and active. In rats fed sub-optimal diets the posterior migration is largely suppressed, which adds support for the second idea. If the backward migration is controlled by the nutritional state of the host, what stimulates the return of the scolex against peristaltic flow to the anterior intestine? The simple answer is that we do not know. A number of attractants have been suggested over the years ranging from bile to 5-hydroxytryptamine (5HT, serotonin) but all have been rejected and the quest continues.

Possibly the most fascinating of all the reports of parasite migration in their host's intestine is that of *Bunoderina encoliae*. *Bunoderina* is a digenean normally found in the rectum of sticklebacks. Four hours after the fish has fed the

Table 5.1. *Competitive exclusion between parasites. In natural infections some parasites are never found infecting hosts in the presence of other species*

Host	Site	Parasite group	Parasite species
Bank vole	Stomach	Nematoda	*Capillaria muris>Mastophorus muris*
	Small intestine	Nematoda	*Heligmosomum halli>Heligmosomum glareoli*
	Small intestine	Cestoda	*Catenotaenia pusilla>Paranoplocephala brevis*
Frog	Lung	Digenea	*Haematoloechus* and *Rhabdias bufonis* are
		Nematoda	mutually exclusive in the same lung
	Small intestine	Protozoa	Only one species of *Opalina* found in single host individual
Rabbit	Small intestine	Cestoda	*Cittotaenia pectinata>Cittotaenia denticulata*
Mole	Small intestine	Digenea	*Itygonimus acreatus>Itygonimus larus*

worms move to the anterior intestine but by 15 hours after feeding they are back in the rectum. In unfed control fish no such migration occurs. The strange feature of this example is that parasites in unfed fish that were in visual contact with feeding fish did migrate, although not as much as in the fed fish. Changes in the intestinal physiology thus seemed to stimulate the parasites even in the absence of food. Shades of Pavlov's dogs!

Other intestinal factors may influence the localisation of parasites. Concurrent infections with the acanthocephalan *Moniliformis dubius* and *H. diminuta* in the rat intestine markedly affect the distribution of the latter. If rats are already infected with *M. dubius* and then infected with the cestode the anteriad ontogenetic migration does not occur and the cestodes establish further posteriorly than normal. When rats already infected with *Hymenolepis* are infected with *M. dubius* the cestodes are displaced posteriorly and the acanthocephalans occupy their preferred site in the anterior intestine. A number of possible explanations have been suggested. There may be competition for attachment sites and mechanical interference, there may be competition for carbohydrate which may become chemically limiting, or it may be that the alcohol produced as an end product of acanthocephalan metabolism may make the anterior intestine less amenable to the cestodes. During these concurrent infections the cestodes appear to get the worst of the deal: not only do they lose their normal homes but are also usually stunted. They do however have the 'last laugh' as, assuming there is no reinfection, *Moniliformis* live for only 6–8 months while *Hymenolepis* may live as long as the host and can eventually return to their own territory. Even among parasites bullies do not get it all their own way.

Concurrent infections may have more extreme consequences than displacement and may result in competitive exclusion. Examples from natural infections demonstrate that this occurs quite commonly (Table 5.1). The factors that control these exclusions have been little studied.

Maintaining station

The site at which a parasite is found is thus influenced by a range of factors, but having established in the best available place it needs to have a way of maintaining its position. To the uninitiated parasites are sometimes regarded as degenerate as they lack some of the organs found in free-living organisms; eyes, after all, are of little use to a parasite living permanently in the dark recesses of its host. But when it comes to attachment parasites are as sophisticated as any animals. The precise attachment requirements depend on the

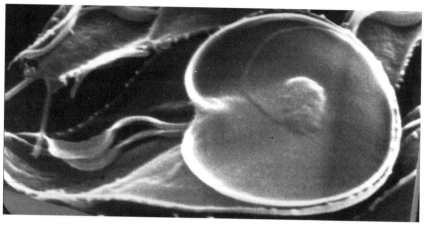

Fig. 5.3 Scanning electron micrograph of the ventral surface of *Giardia* showing the sucker which attaches the protozoan to the host's duodenal cells. From a photograph by Professor K. Vickerman, reproduced with permission.

position in the host that the parasite occupies. In the gut the flow rate in the centre of the lumen is much greater than that at the mucosal surface and to maintain station parasites which occupy the lumen need more powerful attachment methods. Each group of parasites has its own unique methods of attachment. Some are structural and obvious, some biochemical and more subtle and some are downright weird and wonderful.

Intracellular protozoan parasites require no means of attachment once they are *in situ*, but most have at least one extracellular phase in their life-cycle when they may be vulnerable. The molecular cell recognition mechanisms described in the last chapter involve an initial attachment phase before cell invasion to maintain contact between the invader and its target. Most gut protozoans live in the more sheltered regions of their hosts amongst the villi, although *Giardia* (Fig. 5.3) is unusual as its concave ventral surface acts as a pair of suckers which attach to the surface of the host's intestinal cells. In heavy infections the parasite may block the duodenal surface and cause intestinal dysfunction and diarrhoea.

The scolices of cestodes, suckers of digeneans and probosces of acanthocephalans mainly come into the first category of obvious structural attachment organs. The association between the scolex and the intestine may be very intimate (Fig. 5.4) and may contribute to host specificity. If the villi are too far apart the suckers are not able to attach fully and the developing protoscolices are swept away; equally if the villi are too close together the protoscolex is unable to get between them and is again unable to establish. The four suckers

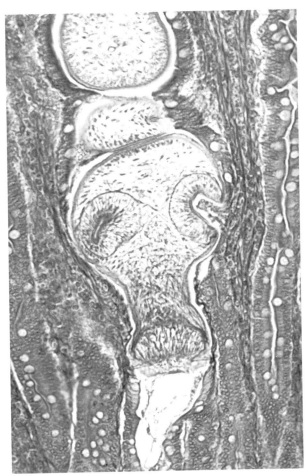

Fig. 5.4 Longitudinal section through the scolex of *Echinococcus granulosus in situ* in the intestine of a dog. Note the extremely close association between the parasite and the host. Host tissue can be seen extending into the sucker on the right. Photograph by Professor J.D. Smyth, reproduced with permission.

and the rostellum of hooks, if present, must make firm contact if the worm is to grow normally. As we have seen *H. diminuta* spends much of its time moving about in its host's intestine and the lack of an armed rostellum may reflect this more nomadic life-style. *Phyllobothrium piriei* (Fig. 5.5), a cestode from the intestine of the ray (*Raja*), comes into the weird and wonderful category. Each individual scolex grows between the villi so that it is uniquely adapted to its host.

There is less variation in the attachment organs of digeneans. Many are

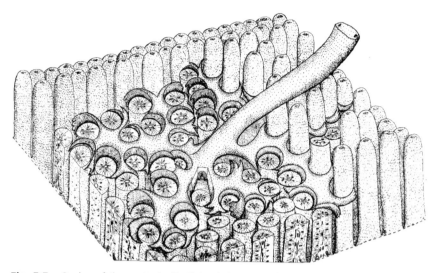

Fig. 5.5 Scolex of the cestode *Phyllobothrium piriei* from the intestine of the ray. Each individual scolex is different as it ramifies between the villi of its host. Adapted from Williams, H.H., McVicar, A.H. & Ralph, R. (1970). *Symposia of the British Society for Parasitology*, **8**, 43–77.

mucosal dwellers and live attached to the gut wall in sites where there is no vigorous flow. Most species have two cup-shaped muscular suckers, an **oral sucker** surrounding the mouth and a second, the **acetabulum**, which varies in position in different species but is normally ventral or posterior (Fig. 5.6). A few species lack one or both but there is comparatively little variation. In the schistosomes the suckers on the male are much more powerful than those on the female, who spends much of her time in the embrace of the **gynecophoric canal** of the male, only emerging to lay her eggs in the finer blood vessels.

Nematodes do not normally have specialised attachment organs and consequently show a greater range of methods of maintaining position than other groups. They tend to be more active than the other helminths and large nematodes such as *Ascaris* are able to hold their position in the lumen of the gut by muscular activity. Some species are able to more or less superficially enter the gut mucosa. *Trichuris* lives with its narrow, whip-like anterior embedded under the epithelium of its host's intestine, while the anterior of *Pseudanisakis rotundata* from the intestine of the ray is more deeply embedded into the submucosa. There is a constriction around the anterior that coincides with the **muscularis mucosa** and helps to lock the worms into place. Some species, including the hookworms, have a large buccal cavity (Fig. 3.6) and, by using

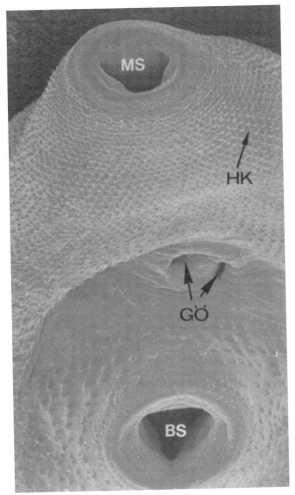

Fig. 5.6 Scanning electron micrograph of the oral (MS) and ventral (BS) suckers of *Fasciola hepatica*. The pore between them is the opening of the reproductive system (GO). From a photograph by Professor Dr H. Melhorn, reproduced with permission.

their powerful oesophageal muscles, they suck in a plug of the gut mucosa which helps to keep them in place to feed on blood released. Hookworms are not permanently attached and change their position regularly. Each time they do so they leave a lesion which continues to bleed, increasing the anaemia which is a consequence of such infections. As we saw earlier *Nippostrongylus brasiliensis* moves in its rat host's intestine with the feeding cycle. The adults also release large amounts of acetylcholinesterase from their **excretory**

Fig. 5.7   Scanning electron micrograph of the anterior of the nematode *Gnathostoma*. Pseudocoelomic pressure inflates the head collar and the spines on it and on the anterior of the body hold the worm in place in the intestine. Photograph by L. Gibbons, CAB International, reproduced with permission.

pores, and it has been suggested that this may inhibit peristalsis locally and thus act as a physiological anchor. Not all nematodes lack attachment organs. *Gnathostoma* spp. from the stomach of cats, dogs, pigs and very occasionally man in various parts of the world have a prominent head bulb (Fig. 5.7) covered with rows of spines. Hydrostatic pressure inside the head bulb causes it to expand and hold the worm in place. This has similarities with attachment by acanthocephalans.

The most distinctive feature of acanthocephalans is the anterior spiny proboscis (Fig. 5.8) which is used for attachment to the host's intestine. In many species the proboscis can be retracted into the proboscis receptacle by muscular action and everted by hydrostatic pressure, but *in vivo* they probably do not move very often. The host often produces a granulomatous reaction to the presence of the worms and this helps to hold them even more firmly in place.

Ectoparasites have as much need for efficient attachment or evasion methods as endoparasites. The forces that they face are probably greater and less predictable and it is thus not surprising that specialised attachment organs

Fig. 5.8   Scanning electron micrograph of the proboscis of an acanthocephalan.
Photograph by Professor Dr H. Melhorn, reproduced with permission.

are a feature of many, but not all. Blood feeding flies, for example, rely on their
speed of reaction to avoid and evade attack, as anyone who has tried to swat
one can testify. Fleas also depend more on speed and their ability to hide
amongst the forest of hairs or feathers that constitute their natural environment to avoid capture. Flies and most species of flea feed for relatively short
periods of time: seconds or minutes rather than hours or days. Some fleas,

ticks and lice are more permanently attached to their hosts and may feed for protracted periods.

Fertilised female *Tunga penetrans*, the 'jigger' or 'chigger' flea from Africa and South America, bury themselves into the skin, often under the toenails, of man where they grow to the size of a small pea, up to 1000 times their initial invasive size, and cause considerable discomfort and distress. The posterior segments which carry the anal and genital openings remain small and protrude through a tiny pore in the skin; from this the eggs are shed. The fleas may get squashed *in situ* and cause even more pain, and there is a risk of tetanus and even gangrene from secondary bacterial invasion.

The recurved teeth on the hypostome of ticks (Fig. 5.9) act in the same way as those on the proboscis of acanthocephalans and hold them in place. Removing ticks is difficult and if care is not taken the hypostome may be left in place to form a focus for bacterial infection. The current recommendation is that ticks should be pulled out as soon as possible after attachment using blunt forceps and the wound treated with an antiseptic.

Lice spend their whole lives on their hosts and their great host and site specificity is partly explained by their means of attachment (Fig. 5.10). The **tarsus** of each leg is provided with a claw which is opposed by a spine on the **tibia**. The size of the aperture between the spine and claw is crucial: if it is either too small or too large the louse cannot obtain an adequate grip on the hair and is likely to be displaced. The distribution of lice on the body is thus dependent on the diameter of the hairs. For example, man may be infected by three species of lice, *Pediculus capitis, P. humanus* and *Phthirus pubis*. The latter has larger tarsal claws and is restricted to the coarser pubic hairs while *P. capitis* is found attached to the finer hairs of the head. *P. humanus,* the body louse, is more commonly associated with clothing rather than living actually on the body. It is very closely related to *P. capitis;* interbreeding may occur and some authors prefer to consider them sub-species.

The mange mites occupy an anomalous position. They are usually included amongst the ectoparasites but many live superficially buried in the skin of their hosts so are at least technically endoparasitic. Both their diminutive size and their habitat make them largely hidden from direct damage from their hosts and as such they have few specialised attachment organs.

The ectoparasitic monogeneans are mainly skin or gill parasites of fish. The skin parasites have an opisthaptor which consists of a single disc provided with 1–3 pairs of hooks and often marginal hooklets. As fish usually swim forwards the water flow over the surface is largely predictable and the worms attach with the haptor pointing into the prevailing flow so that they offer the least resistance. Many, in addition, have adhesive glands whose sticky secre-

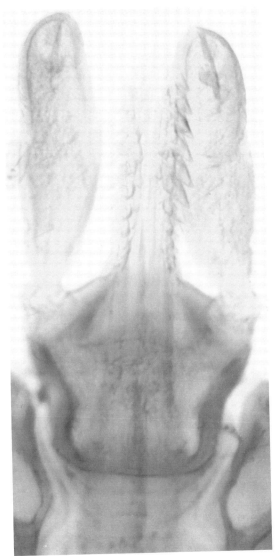

Fig. 5.9 Mouthparts of a hard tick. Touch sensitive pedipalps border the toothed hypostome that holds the tick in place and through which the cutting chelicerae extend.

tions help to hold the anterior to the skin surface. Within gill chambers the water flow is more turbulent and the haptor of gill parasites is modified to provide a number of suckers or clamps around its edge which attach to individual gill filaments (Fig. 5.11). The haptor may grow asymmetrically to allow

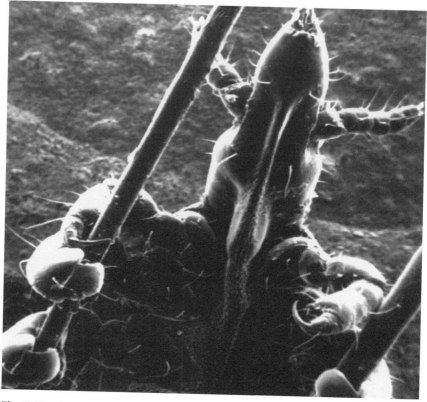

Fig. 5.10 Anterior of a sucking louse attached to the hairs of its host by the tarsal claws. Photograph by Dr R. N. Titchenor, reproduced with permission.

better adhesion and analysis of records from natural infections suggests that the adaptations are highly specific (Table 5.2).

Resisting host attack

All organisms, free-living or parasitic, must have the basic ability to resist the normal conditions of their environment. The outer surface is the first line of defence against hostile conditions and must be up to the task if the organism is to survive. For cestodes and acanthocephalans, which lack a functional gut, the body surface has to be able to fulfil the dual roles of absorption and protection while the outer surface of nematodes, monogeneans and digeneans, which do have guts, is largely protective, although some nutrient transfer may also occur.

The tegument of the different groups of platyhelminths shows similarities

Table 5.2. *Distribution of monogeneans on the gills of naturally infected whiting*

| Species | Whiting gill arch | | | |
	I	II	III	IV
Diclydophora merlangi	53	8	1	4
Diclydophora bissae	7	118	91	8
Corthocotyle merlucei	4	28	7	4

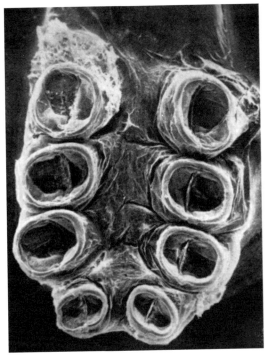

Fig. 5.11 Opisthaptor of a monogenean, *Neodiscocotyle carpioditis*, from the gills of a quillback, *Carpiodes cyprinus*. The normal suckers are modified into clamps that are better able to withstand the multidirectional water currents that occur over the gill filaments. Photograph by Wright, K.A. & Dechtiar, A. (1974). *Canadian Journal of Zoology*, **52**, 183–187.

that testify to a common ancestry (Fig. 5.12). In essence it consists of a number of layers: a metabolically active outer, anucleate, syncytial layer with the nuclei of the syncytium forming an inner, nucleated layer embedded in the parenchyma and separated by layers of circular and longitudinal muscle. The

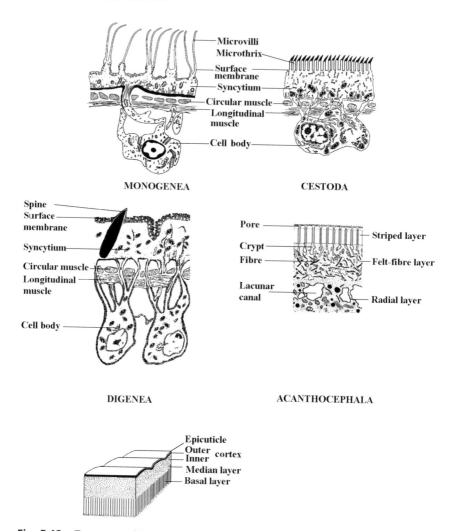

Fig. 5.12 Tegument of helminths. Similarities are apparent in the syncytial teguments of platyhelminths, while those of acanthocephalans and nematodes are distinctive.

outer surface membrane is covered by a thin (up to 40 nm) **glycocalyx** which shows up in electron micrographs as a 'fuzzy coat'. It consists largely of mucopolysaccharides and glycoproteins and has a net negative charge so that it can bind organic and inorganic ions. Host enzymes may be adsorbed to the glycocalyx so that they may be used by the parasite to give it an advantage over the host that synthesised them. The glycocalyx is continuously replaced and may also have a role in protection against other host enzymes and immune responses.

The outer layer differs in the different groups. Both the Monogenea and Cestoda have microvilli on their outer surface, although these differ in detail. They are best developed in the Cestoda where they are known as microtriches (singular microthrix) and form a more or less continuous covering; there are fewer of them on the scolex and neck. Each microthrix is covered by an electron-dense cap. The effect of the microtriches is to increase the surface area by up to ten times. They have been compared with the microvilli on intestinal cells and are considered to be similarly associated with nutrient absorption. The microvilli on monogeneans lack an electron-dense cap and tend to be longer and less regularly arranged than those on cestodes. The increase in surface area is, at most, only about twice, and some monogeneans lack them completely. This suggests that they may have alternative functions and several, including mixing sticky secretions to hold the worm in place and providing support for a protective layer of mucus, have been suggested in addition to the normal nutrient and gaseous exchange functions. The digeneans lack microvilli but are typically covered by spines. These occur more densely anteriorly and are mainly associated with maintaining position in the host.

Some absorption occurs across the tegument of all platyhelminths. For monogeneans and digeneans this tends to be restricted to small molecules and their teguments may be folded and have crypts to increase the absorptive area. In addition the tegument is secretory although relatively little is known about the precise secretory products. Like the platyhelminth tegument, that of acanthocephalans is syncytial and coated with a glycocalyx, but there the similarities end. The acanthocephalan tegument consists essentially of three layers. The outer surface is pierced by numerous fine pores, 15 nm in diameter and about 200 nm apart. These lead into crypts that increase the surface area up to 60 times, have enzymes associated with them and are normally interpreted as channels through which nutrients may be absorbed by **pinocytosis**. This outer layer is termed the **striped layer** and is bounded on its inner side by a thicker **felt-fibre layer** typified by large numbers of collagenous fibres that run in all directions. The thickest layer is the inner, **radial layer** which also contains fibres and like the felt-fibre layer is structural, but in addition contains nuclei, mitochondria and the main channels of the lacunar system, and probably constitutes the centre of metabolism for the worm.

The body wall of nematodes generally consists of three regions, muscle, **hypodermis** and **cuticle** and differs from the other helminths in that it is not normally the seat of nutrient absorption. The exception that proves this rule is found amongst some insect parasitic species, for example, adult female *Bradynema* sp. live in the haemocoel of the maggots of the dipteran mushroom pest *Megaselia halterata*. Once in their host they lose their feeding apparatus and develop microvilli on their surface through which it is assumed they obtain all their nutrient requirements. For the rest of the nematodes the cuticle is secreted by the hypodermis and has three layers, an inner basal, middle median and outer cortical layer. Different species have different arrangements and some layers may be reduced or missing. The cuticle is largely structural and, although there is some evidence that some small molecules may be able to cross it, mainly acts to withstand the high pseudocoelomic pressure that develops during nematode locomotion and to protect the worms from environmental conditions and host digestive enzymes. While the worm is alive and the cuticle intact it is safe from attack but on death or with a damaged cuticle it is rapidly digested. Some large nematodes, e.g. *Ascaris lumbricoides* also produce antienzymes which may inhibit the host's digestion and also increase protection for the worm.

Blood parasites face a different set of hazards. Rather than hostile host enzymes they encounter the full force of the host's immune system (see Chapter 6). Intracellular blood protozoans can hide within host cells and thus largely escape notice, although they are, of course, vulnerable when they change cells. Extracellular blood dwellers need alternative methods if they are not to be eliminated before they have even established. Trypanosomes, for example, are continually changing the glycoprotein molecules (**variable surface glycoproteins, VSGs**) that coat the outer surface of the plasma membrane. As soon as the host has produced antibodies to destroy the population of trypanosomes with one VSG another population develops with a different VSG and the host is back where it started, making another set of antibodies. Trypanosome infections are characterised by cyclical fluctuations in **parasitaemia** (Fig. 5.13) with the host desperately trying to catch up. The body is in much the same situation as William Gladstone, the 19th century British prime minister who, according to Seller and Yeatman in '1066 and All That', 'spent his declining years trying to guess the answer to the Irish Question. Unfortunately whenever he was getting warm the Irish secretly changed the question'.

Schistosomes also live in the blood stream and the host is able to mount an immune response against them. This response is only effective, however, against the invading stages of a further infection and the existing worms are unaffected. This **concomitant immunity** in which established worms can live in an otherwise immune host is a consequence of the incorporation of host molecules into the tegument of the worms early in the primary infection, before the host has had a chance to build up antibodies against them. These disguised worms are recognised as 'self' by the host and left in peace, while invading schistosomula are attacked and destroyed on entry. Schistosomes

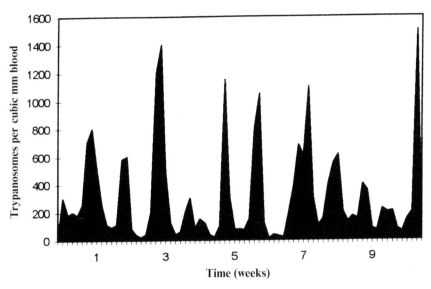

Fig. 5.13 Variation in parasitaemia during infection with *Trypanosoma brucei.* Each peak represents a population of parasites with a different variable surface glycoprotein. Redrawn from Ross, R. & Thompson, D. (1910). *Proceedings of the Royal Society of London,* **B82**, 411–415.

may be long lived and concomitant immunity allows the established adults to continue egg production for long periods without competition from new invaders, and may also prevent the host from becoming over-burdened by a potentially lethal infection.

Obtaining nutrients

As the nematode cuticle is not absorptive nematodes rely almost exclusively on their gut for nutrient uptake. The intestinal wall is only a single cell thick. The inner surface is microvillous, is both secretory and absorptive, and a range of enzymes has been found associated with it. Fluid food is pumped into the intestine by the oesophageal muscles and after digestion and absorption is voided through the anus. To ensure the correct consistency of the food ingested some tissue feeders, e.g. *Chabertia ovina* (Fig. 5.14) from the colon of sheep and goats, pre-process their food by extracorporeal digestion inside their enormous buccal capsules, utilising enzymes released from the oesophageal glands. In these species the intestinal cells tend to be largely absorptive leaving the secretory function to the oesophageal glands.

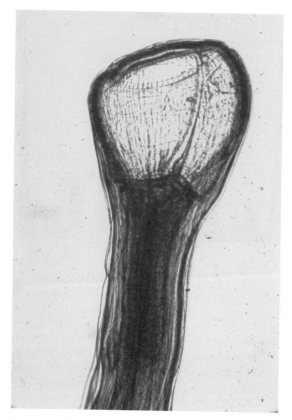

Fig. 5.14 Buccal capsule of the nematode *Chabertia ovina*. A plug of gut mucosa is sucked into this and secretions from the oesophageal glands initiate extracorporeal digestion.

Both the Monogenea and Digenea also have functional guts, although they normally lack an anus. Relatively little is known about the detailed feeding in either group. The monogenean *Diclydophora merlangi* has been shown to feed on blood taken from the gills of its whiting host. Initial breakdown of the blood occurs in response to enzymes produced by glands in the pharyngeal region so that when it reaches the gut caeca it has a homogeneous, acellular consistency. The haemoglobin is absorbed into the caecal cells by endocytosis, broken down by lytic enzymes inside the cell and the indigestible haematin residue discarded into the caecal lumen by **exocytosis**. Digestion is thus largely intracellular.

Our knowledge of feeding in digeneans is based largely on detailed studies of the liver fluke *Fasciola hepatica* and the blood dwelling *Schistosoma mansoni*.

Schistosomes live in and feed on blood. Proteolytic enzymes in the gut lumen break down the haemoglobin to produce small peptides which are absorbed. The iron-rich undigested waste products are then regurgitated. Questions are still raised about the relative importance of blood and host liver tissue cells in the diet of *Fasciola*. However, it seems unlikely that the worm selectively avoids one cell type and it is much more likely that both contribute to the daily food intake. In the caeca the cells pass through a cycle of secretion and absorption. If food is absent there is a build-up of secretory bodies which pass to the apical portions of the cells and are released when food enters the gut. After release of the enzymes into the lumen the cells change their function and absorb the amino acids and peptides that have resulted from the digestive process. Unlike the Monogenea, digenean digestion is largely extracellular. The fact that the only two species that have been examined in any depth have such different solutions to essentially the same problem suggests that we still have an enormous amount to learn about helminth nutrition and digestion.

Nutrient requirements

It is easy to say and often said that parasites rely on their hosts for all their daily needs, but what exactly are those requirements? There are as many answers to that question as there are parasites and exact details are still not known. *In vitro* culture techniques have resolved some issues but even protozoans are much more complex than bacteria and most culture media contain such ingredients as serum or yeast extract, the precise chemical composition of which is not known. Culture in chemically defined media has rarely been achieved. It is, however, possible to make some generalisations and to offer a few specific examples.

Like any other animal, parasites require a source of energy and an environment that is suitable to exploit it. At one time it was considered that gastrointestinal parasites lived in an anaerobic environment and thus required exclusively anaerobic metabolic pathways. More recent measurements suggest that even in the centre of the intestinal lumen there is normally some oxygen and that this increases as the mucosal surface is approached. Truly anoxic conditions may obtain in the centre of the faecal mass as water is absorbed from it in the large intestine, but most gastro-intestinal parasites have access to some oxygen. It is thus not surprising to discover that many parasites have both aerobic and anaerobic pathways available to them and can switch as conditions dictate.

Typically parasites metabolise glucose or glycogen using anaerobic glycolytic pathways. Unlike vertebrates, where glycolysis is used mainly as an emergency measure for rapid muscular activity, parasites rely on it for much of their carbohydrate catabolism. In vertebrates the end product is normally lactate but for parasites there is a much wider range of comparatively energy-rich end products which are excreted and may subsequently be used by the host (Fig. 5.15). These pathways tend to be relatively less energy-efficient than their equivalents in mammals, but this loss in efficiency may be compensated for by the plentiful supply of substrates provided by the host.

If 'man shall not live by bread alone' nor shall parasites exist solely on carbohydrate. Many other nutrients may be absorbed and utilised. Lipids are absorbed and stored and used for production of cell membranes which, as they often form the host–parasite interface, are especially important and constantly replaced. For many animals lipids also provide a valuable energy source, especially during periods of starvation. Most parasites lack the necessary enzymes but some non-feeding, infective, third stage nematode larvae do use lipid stored as a result of feeding by the first and second stage larvae as their major energy source.

All parasites can synthesise many of the amino acids they require to make enzymes and structural proteins. They often use CO_2 fixation to achieve it. *In vitro* studies have pinpointed some examples where such *de novo* synthesis is not possible, and in these cases the parasite concerned must obtain the amino acid from its host.

Purines and **pyrimidines** are essential for the production of the nucleic acids DNA and RNA, and are also involved in many other metabolic reactions. Many parasites reproduce very rapidly and their need for DNA is thus especially great. Present evidence suggests that, while most parasites can synthesise pyrimidines, few, if any, can synthesise the purine ring. Purines must thus be obtained from the host and salvage pathways exist to allow the parasite to recycle the host's molecules.

Relatively little is known about specific nutritional requirements of individual parasites and many no doubt have them, as the two examples below illustrate.

In mammals tetrahydrofolate is essential for cell metabolism and especially for the production of red blood cells in the bone marrow; a deficiency is characterised by anaemia, largely due to interruption of normal purine and pyrimidine synthesis. Mammals normally produce tetrahydrofolate from folic acid, one of the B vitamins. The malaria parasite, *Plasmodium*, is unable to utilise this pathway but can synthesise tetrahydrofolate from simpler precursors including *p*-amino benzoic acid (PABA). Mice maintained on a milk

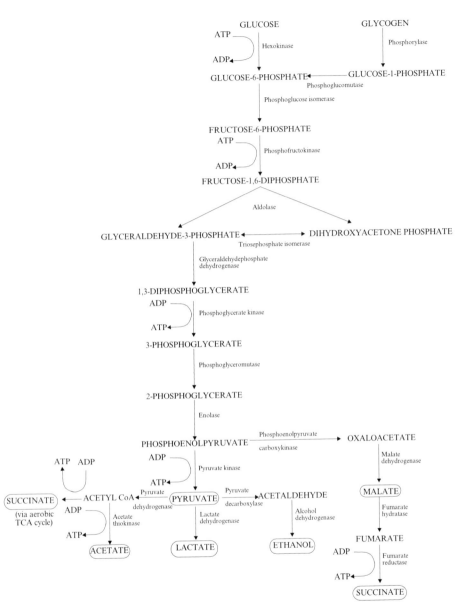

Fig. 5.15 Glycolytic pathways and end products of parasite metabolism.

diet which is very low in PABA are largely refractory to *Plasmodium* and laboratories that passage the parasite through mice for experimental purposes supplement the normal mouse diet with PABA in the drinking water.

The large pseudophyllidean cestode, *Diphyllobothrium latum*, has an insatiable appetite for another B vitamin, B_{12}, cobalamin. In humans a lack of vitamin B_{12} results in a condition known as **pernicious anaemia** which is marked by a disproportionately large number of large, immature erythrocytes in the circulation. The maturation of these cells has been prevented by a lack of vitamin B_{12}. Normally the body gets all the cobalamin it needs from the diet. The vitamin forms a complex with a component of the gastric juice known as **intrinsic factor**, and in this state it is absorbed in the intestine and stored in the liver. Patients who produce too little intrinsic factor are liable to develop pernicious anaemia. Uniquely, the enormous appetite the *D. latum* has for vitamin B_{12} means that it competes with its host for the available vitamin and in most cases wins. It is not yet clear whether the tapeworm has especially high affinity receptors for the vitamin, whether it produces its own intrinsic factor-like molecule which is more attractive than that of the host, or whether it produces an enzyme that cleaves the normal complex, releasing the vitamin. Whatever the mechanism it is highly efficient and if radiolabelled vitamin B_{12} is fed to tapeworm carriers between 80 and 100% of the label is recovered in the worm. As only 1–2 in 1000 *D. latum* cases end in pernicious anaemia, the amount of the vitamin stored in the liver may be quite high and turnover relatively slow, so that it is not an inevitable consequence of infection. Removal of the worm by **chemotherapy** almost always results in complete remission of the anaemia.

This last example illustrates an important point. Cestodes and acanthocephalans especially are in competition with their hosts for nutrients in the gut lumen. If they are to compete successfully their absorptive ability must be at least as efficient as that of their host or, if there is any shortage, they will be out-competed and will starve.

6

The host's response

So far we have mainly considered parasites as specialised animals, with many of the attributes of free-living relatives, that happen to inhabit an environment in or on another animal. The host has been relegated to the role of 'black box' providing the essential requisites for the parasites' health and welfare. Parasites are, however, unique as they are the only animals, other than man, whose whole environment changes as a direct result of their presence, and the host/parasite relationship is a dynamic interaction in which both participants evolve together. The working definition of a parasite introduced in Chapter 1 incorporated the idea that a parasite necessarily causes damage to its host, and we will look in more detail at specific types of damage in the next chapter, but the host has ways to reduce any damaging effects of parasites and it is these that will now be addressed.

The evolutionary pressures acting on host and parasite have been likened to an arms race between feuding nations. Not the sort of arms race where the protagonists vie with one another for ever larger and more destructive weapons but rather the situation where one side makes a larger warhead and the other prepares a better shield to defend itself – shades of Star Wars, the United States' Strategic Defense Initiative of the late 1980s rather than the film, although that may not be totally inappropriate. Even evolutionarily long-established host/parasite pairs still exhibit pathological consequences and it has been argued that the pathology must benefit one of the participants or natural selection would have led to its exclusion and asymptomatic infections. From theoretical work with mathematical models it has been argued that natural selection will act to maximise parasite reproductive success. If, as seems probable, transmission is linked to pathogenicity then selection will lead to high to intermediate **virulence**, the situation that we generally see.

As hosts have evolved methods of controlling their parasites, so the parasites have evolved means of circumventing them, and we saw one or two examples of these in the last chapter. The generation time of a parasite is generally substantially shorter than that of its host and its fecundity much greater so that the rate of evolutionary change is likely to be much greater for the parasite than the host. On the other hand any 'evolutionary error' on behalf of the parasite is likely to have devastating consequences. If the parasite fails to establish, or is rapidly overcome and thus fails to reproduce, it contributes nothing to the next generation and is extinct, whereas if the host becomes more susceptible it may suffer some additional discomfort and disease but this is unlikely to result in a total failure to breed. Over time a sort of 'armed truce' has arisen with both host and parasite tolerating each other but always ready to exploit any weakness in the other's armour. Although many parasites have the potential to cause serious disease (see Chapter 7) in natural infections relatively few do so and both host and parasite populations are held in a dynamic balance. Like any other balance this one may be thrown out of equilibrium. For example, the protozoan parasites *Toxoplasma* and *Cryptosporidium* cause mild, self-limiting infections in most immunocompetent people, although *Toxoplasma* may have more serious consequences for pregnant women, but in AIDS patients and others whose immune system has been compromised, both parasites may cause life threatening conditions. It is perhaps pertinent to point out here that many natural parasite infections in animals are well tolerated and leave few signs of their presence, although they have the potential for causing serious damage if the parasite population is too great. Often the factor that results in these large, damaging populations is man's intervention.

The immune system

Any organism that successfully breaches the body's outer defences, the skin, acid in the stomach, mucus in the intestine, etc., is faced by the might of the immune system. The vertebrate, and especially the mammalian, immune system is immensely complex. It depends upon an individual being able to distinguish between molecules that originate endogenously, 'self', and those of foreign origin, 'non-self', and to react appropriately to the latter. 'Self' recognition depends on specific molecules, known as **major histocompatibility complex** (MHC) molecules on the cell surface of each cell in the body.

The majority of cells have class I MHC molecules but some specialised cells, including **macrophages** and some **B lymphocytes** have class II MHC molecules and these indicate alternative properties. This recognition of 'self' and 'non-self' is innate and the body's immune system 'learns' to recognise its own cells very early in foetal life.

Bacteria or viruses that manage to enter the blood and tissues are usually rapidly recognised as foreign, engulfed by voracious phagocytic macrophages, digested and destroyed. Phagocytosis constitutes one of the major processes in the innate or natural immune system, is a first internal line of defence and does not rely on any prior experience of the invader. Macrophages are too small to ingest most animal parasites although they may try hard to do so, and, while there are many factors that contribute to it, natural immunity to parasites is more or less equivalent to host specificity, which we have already met. Other enzymatic and chemical components of the natural immune system which are effective against bacteria and viruses are less significant against animal parasites, although they may have an indirect effect on the course of some infections. For example, **transferrin**, a globulin circulating in the blood with an avid appetite for iron, may out-compete the malaria parasite for limited iron stocks. A mild anaemia may actually be advantageous to the malaria patient and famine victims are sometimes vulnerable to infections that remain latent during times of privation, only to manifest themselves when relief raises the plane of nutrition so that the parasite can obtain adequate iron.

If several hundred infective third stage larvae of the nematode *Nippostrongylus brasiliensis* are placed on the skin of naïve rats that have never seen the parasite before they migrate through the skin and then pass in the circulatory system to the lungs and subsequently to the gut lumen, where the adult worms mature and begin to release eggs about five days after infection. Egg production reaches a peak about day 9 and thereafter there is a rapid decline until by about day 21 post infection it has ceased and the adult worms have been expelled from the intestine (Fig. 6.1). The rats are now largely refractory to reinfection and even if a massive challenge dose 5–10 times the original is administered few adults mature, relatively few eggs are produced and the infection is eliminated within about ten days. The rats have acquired resistance and are immune from further infection. The ability of the body to recognise and react more rapidly and more intensively to 'non-self' molecules that it has seen before is the main consequence of the specific acquired immune responses which constitute the body's other major defence mechanism and depends on the cellular and chemical interaction of innumerable lymphocytes.

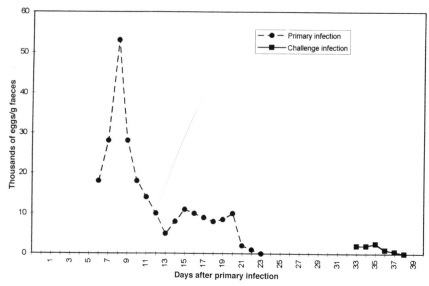

Fig. 6.1 Course of primary and challenge *Nippostrongylus brasiliensis* infections in the rat. 100 larvae were administered on day 0 of the primary infection and egg release was monitored until the infection was eliminated on day 23. The rats were reinfected with 500 larvae on day 30 and again egg release was monitored. Data from Africa, C.M. (1931). *Journal of Parasitology*, **18**, 1–13.

Lymphocytes are responsible for both the recognition of antigens and taking appropriate action against them. To achieve these various functions there are two or three major and, as each individual **epitope** may have its own population of specific lymphocytes, countless minor populations of lymphocytes in the body. All lymphocytes originate from stem cells in the bone marrow and the two major populations differ in their subsequent processing (Fig. 6.2). B-lymphocytes (B-cells) are processed in the **Bursa of Fabricius** in birds or the bone marrow in mammals and, on maturation, produce **immunoglobulins (antibodies)** which react with antigens to neutralise and render them harmless to the host. **T-lymphocytes (T-cells)** are processed in the thymus gland and differentiate into two distinct subsets, **T helper (Th) cells** and **cytotoxic T-cells (Tc).** T helper cells have a molecular marker on their surface designated CD4, while cytotoxic T-cells are recognised as having a CD8 molecular marker. Both B-cells and T-cells pass to the lymph nodes and the spleen so that they may be available to react to antigens carried in the blood or lymph.

Large foreign protein and carbohydrate molecules (**antigens**) which enter the body are recognised by macrophages and a complex chain of reactions ensues that eventually results in the control or elimination of the intruder.

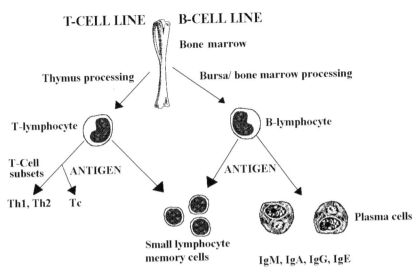

Fig. 6.2 Processing of lymphocytes produced in the bone marrow to give either T-cells or B-cells. In the presence of antigen the former are associated with cell mediated immunity while the latter produce antibodies (immunoglobulins) and are concerned with humoral immunity. Both types may transform into memory cells that remain dormant in the body to respond more rapidly to subsequent exposure to the same antigen.

Rather than recognise the whole antigen molecule only specific, immunologically active three-dimensional structural arrangements (epitopes) are recognised. Thus a large helminth may present many complex antigen molecules, each with a number of epitopes, to the immune system. Like a newly appointed ambassador, antigens must be formally and correctly presented if they are to be recognised and responded to. On entry into the body antigens are recognised by macrophages as 'non-self' by their lack of compatible MHC molecules and ingested. Enzymatic degradation of the large antigen molecule occurs intracellularly and the individual epitopes, peptides of some ten amino acids in length, combine with internal class II MHC molecules and the complex passes to the macrophage surface and projects through it. The macrophage is thus transformed into an **antigen presenting cell** (APC) (Fig. 6.3).

In this state the epitope can be recognised by one of the vast number of CD4+ T-lymphocytes that have been waiting for just this moment. During early embryonic life the immune system is busy producing T-cells with specific T-cell receptors (TCR) for each and every possible epitope. The system is supremely sensitive: peptides that differ by only a single amino acid are recognised as different. We thus come into the world primed for encounters

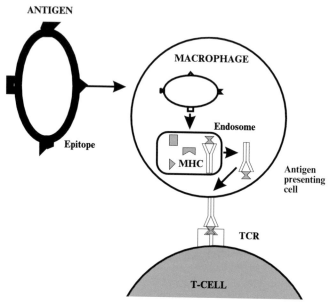

ANTIGEN

MACROPHAGE

Endosome

Epitope

MHC

Antigen
presenting
cell

TCR

T-CELL

Fig. 6.3 Antigen recognition. Foreign antigens are phagocytosed by macrophages, and individual epitopes are stripped off and then presented on the surface of the cell to be recognised by a complementary T-cell.

with both likely and unlikely antigens. Many of these 'bridesmaids' will never meet their allotted partner but those that do, under the influence of a chemical **cytokine**, interleukin 1 (IL-1), undergo cell division to produce a clone of identical lymphocytes all with the same TCR. Some of these return to the resting state as small memory cells, of which more later, and others release cytokines that stimulate other cells of the immune system to react against the antigen.

Cytokines are low molecular weight proteins that control the activities of the immune system. They consist of a dozen or so **interleukins, interferon** (IFN), **tumour necrosis factor** (TNF) and **colony stimulating factor** (CSF). Their interrelationships are complicated (Fig. 6.4) but their presence results in the switching on or off, or in immunological speak, up-regulating and down-regulating, of other parts of the immune system.

As well as entering the body from outside antigens may occur endogenously, for example from intracellular parasites. These are enzymatically digested into peptides, transported to the endoplasmic reticulum where they combine with class I MHC molecules, and again the epitope/MHC complex is presented at the cell surface where it is recognised by CD8+ cytotoxic T-cells. As most cells in the body express MHC class I molecules they all have the potential to present antigen to CD8+ T-cells this way.

Once an antigen has been recognised, correctly presented to the T-cells and

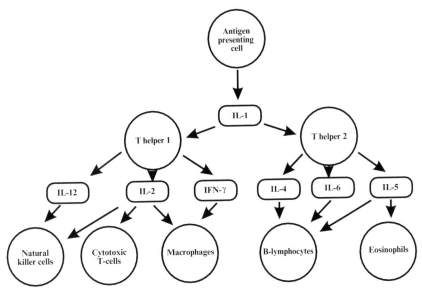

Fig. 6.4 Interactions between cytokines and the cells of the immune system. Each of the immune processes is stimulated or inhibited by cytokines which act as chemical control factors.

a clone of complementary T-cells established, the effector side of the immune system can come into play. Effector mechanisms fall into two broad categories, humoral and cell mediated. Activated T-cells also occur in two subsets, Th1 and Th2, depending on the cytokines they secrete (Fig. 6.4). In general Th1 cells produce IL-2, IL-12 and interferon which activate macrophages, natural killer cells and cell mediated responses, while Th2 cells release IL-4, IL-5 and IL-6 which stimulate B-cells to produce antibody (immunoglobulin) and initiate humoral responses.

Instead of a TCR to recognise a specific epitope each B-cell is programmed to produce a single antibody. This is presented on its surface membrane and acts as the receptor for its matching antigen. B-cells are triggered to grow and multiply by the presence of antigen, IL-4, IL-5 and IL-6. Some of the daughter cells remain large, differentiate and mature into antibody-secreting **plasma cells** while others become small **memory cells**. Following an initial infection it takes time for the immune system to mobilise its resources and begin to produce antibody: several days elapse before significant amounts appear. After a subsequent infection the bank of memory cells can be brought immediately into action and the response is both faster and more intense.

Antibodies are proteins, immunoglobulins, that can recognise antigens and initiate reactions against them. They are composed of four peptides: two,

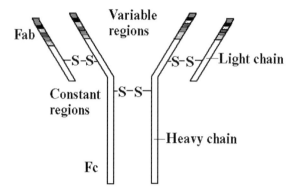

Fig. 6.5 Generalised structure of an antibody. The crystallisable Fc region is constant within any class of antibody but the Fab region is highly variable and it is this region that is specific for each possible individual antigen.

known as the heavy chains, about 400 amino acids long and two, the light chains, about half that length, arranged in a Y shape (Fig. 6.5). The arms of the Y contain highly variable regions on both the light and the heavy chains and it is this (Fab) region that binds to the complementary antigen. The stem of the Y consists of a crystallisable fraction, the Fc region, which provides a link with the rest of the body. It differs between different antibody classes but is constant within each class.

Each antibody thus has a recognition site, the Fab region, and an effector site, the Fc region. Antibodies can initiate a number of immune reactions. By coating the invader they make it more attractive to macrophages which have receptors for the Fc portion. These activated macrophages, which possess higher proteolytic and other cytolytic activities, can damage or destroy relatively large parasites in a process known as **antibody dependent cell mediated cytotoxicity** (ADCC to its friends).

In addition antigen/antibody complexes can activate (fix) the cascade of reactions that constitute the classical **complement** pathway (Fig. 6.6). The complement system consists of a complex collection of serum proteins which, once triggered, react sequentially to eventually produce an enzymatically active **membrane attack complex** (MAC) which can be directed against the source of the antigen. Each step in the sequence requires the presence of the products of the previous reaction as enzymatic catalysts and the process is amplified as it proceeds (designated by the letter b in Fig. 6.6). The final MAC increases the permeability of cell walls so that they are destroyed by osmotic stress. Additionally components of the complement system attract macrophages, enhance phagocytosis and increase local vascular permeability. A

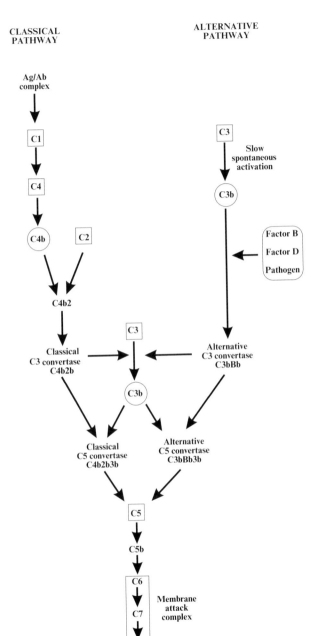

Fig. 6.6 Classical and alternative complement pathways. Once triggered the pathways produce biochemically active products, the Membrane Attack Complex, which damage cell walls.

number of control mechanisms prevent the system running amok and lysing healthy body cells. While it is the antigen/antibody complex that initiates the classical complement pathway there is an additional mechanism, known as the alternative complement pathway, that is triggered by some micro-organisms, including some parasites, that does not require the intervention of antibodies.

There are five classes of immunoglobulin (Ig), four of which are significant in parasite infections. The fifth, IgD, appears to be associated with receptor functions on B-lymphocyte cell membranes. The first to appear after exposure to antigen is IgM. This consists of five Ig molecules bound together. In addition to enhancing phagocytosis and fixing complement the ten antibody binding sites make it an effective agglutinator, causing cells to clump together. It is secreted across mucosal surfaces so that intestinal parasites are exposed to it. IgM molecules last for 4–5 days, after which time they are replaced by the most abundant of the immunoglobulins, IgG.

Immunoglobulin G exists in a number of subclasses, can cross the placenta and thus transfer maternal immunity to the foetus and has broadly similar immunological properties to IgM, although it is not secreted across mucosal barriers and is more likely to be involved in ADCC. Immunoglobulin A is readily secreted in many body fluids including tears, saliva and milk. It is the component of mothers' milk that helps to boost transplacental immunity and confer protection to the newborn. It is synthesised at mucosal surfaces to prevent their breach, is stable to enzymatic breakdown in the gut and, along with IgM, is thus available against intestinal parasites and can trigger the complement pathway by the alternative route. Immunoglobulin E, unlike the others, is not free in the serum but binds to **mast cells** and **basophils**. In the presence of antigen and IgE the mast cells degranulate, releasing biologically active amines including **histamine** (Fig. 6.7). This process, known as **immediate hypersensitivity** (IH), is responsible for allergic reactions such as hay fever and asthma and is especially significant in helminth infections. IgE also facilitates ADCC.

Immune response to parasites

We are now in a position to return to the *Nippostrongylus* example introduced earlier and consider what may be happening inside the host following primary and challenge infections. Initial contact with the immune system occurs as the larvae penetrate the dermal tissue in the skin to enter the blood. The larvae are large compared to macrophages, nematode cuticle is notoriously unreactive and translocation from the skin to the lungs takes less than 24 hours. These factors mean that although macrophages, T-cells and B-cells may all be mobil-

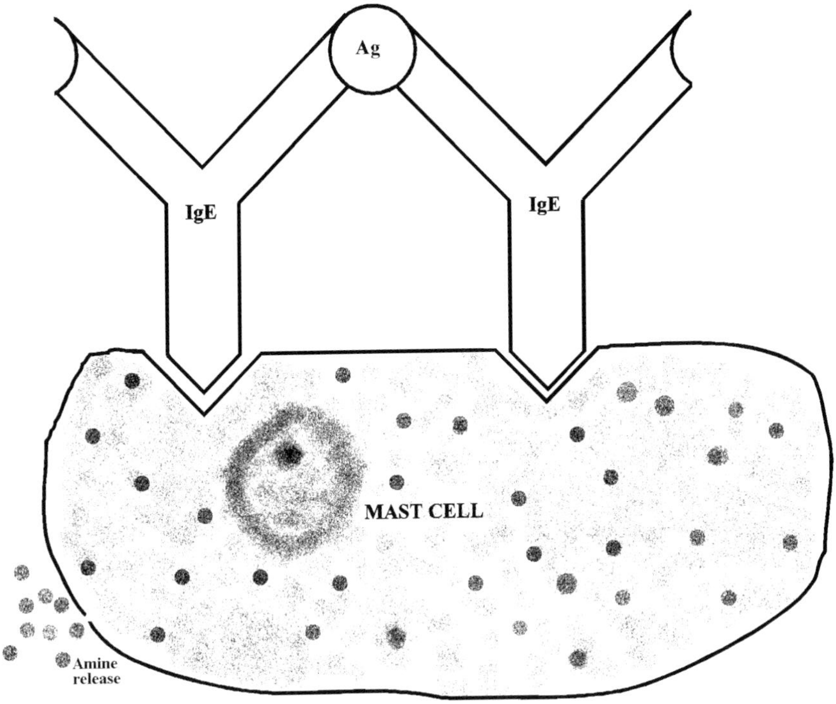

Fig. 6.7 Release of biochemically active amines from mast cells. Two IgE molecules linked with antigen are required to fit into complementary receptors on the mast cell surface to initiate amine release.

ised they have little chance to mount an effective response. In the lungs the third stage larvae moult and the fourth stage larvae (L4) present a different set of stage specific antigens. Within another two days the L4 have ascended the ciliary escalator and passed into the intestine where the final moult occurs with yet another series of surface antigens to challenge the immune system. The lag time between exposure and effective response is thus exploited by not presenting the same antigens for long enough for the host to react against them.

In the gut the adults live between the villi where they are exposed to IgA at the mucosal surface, but most significantly to IgE which is produced in large quantities in helminth infections and contributes to immediate hypersensitivity reactions that eliminate the adult worms. The exact mechanism of expulsion is still not fully understood. The amines released by mast cell degranulation in the presence of IgE and worm antibody may directly damage the worms, the inflammation caused in the intestinal mucosa may make the normal limited site of infection untenable by the adults, and changes in local vascular permeability may release specific anti-worm IgG that attacks the worms. The overall effect is that the worms are expelled from the rat. On challenge infection the memory cells that were formed in response to the primary infection are rapidly mobilised and become effective early enough to participate in ADCC against both third and fourth stage larvae as they pass through the circulation, reducing the percentage that reach the intestine. The protective measures in the gut again become effective more rapidly and more intensively to result in decreased egg production and rapid worm expulsion.

The 'spontaneous' ejection of gastro-intestinal nematodes, sometimes known as the 'self-cure' phenomenon, also occurs in some economically important species, for example *Haemonchus contortus,* a serious pathogen of ruminants. However, it is a feature of most parasite infections that they are long term and chronic, although the parasites themselves are highly **immunogenic**. With so many host defences against them the surprising thing is that so many parasites can live in such a hostile environment. In the last chapter we saw how *Trypanosoma* stays one step ahead of the immune system by continually changing its antigenically active surface coat and how schistosomes remain immunologically invisible by incorporating host molecules into their tegument so that they are recognised as 'self' rather than 'non-self' and other species have other ways of immune avoidance.

If, instead of being exposed to large primary and challenge infections, rats are infected daily with only a few *Nippostrongylus* larvae, the infection may be extended for many weeks (Fig. 6.8). This **'trickle infection'** may be a better reflection of many natural infections. The exposure to the parasite remains below the host's immunological threshold so that relatively little protective response occurs against it.

Some parasites live in parts of the body where they are protected from the immune response. Metacercarial stages of digeneans are usually surrounded by a thick protective cyst wall, but those of *Diplostomum spathaceum,* a digenean from the gut of gulls, live in the lens of the eye of their fish intermediate hosts. The lens is an 'immunologically privileged' site where the immune system is not effective so that additional protection is not required.

Intracellular parasites are also largely separated from the direct effects of immune attack, but this does not mean that such infections do not invoke protective responses. Malaria is a disease which affects some 200 million people and causes more that 1 million deaths each year. Most of these deaths occur among children, adults in highly malarious regions survive with the infection. Because of the enormous multiplicative potential of the malaria life-cycle (Fig. 2.6) infection with a single sporozoite can theoretically result in an overwhelming infection and, if there was no protection, could have resulted in the total depopulation of the vast areas liable to malaria transmission. Protection offered by the Duffy antigen was mentioned in Chapter 4 and other natural pro-

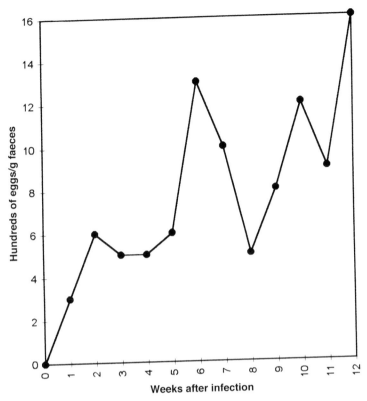

Fig. 6.8 Egg release from rats infected with five *Nippostrongylus brasiliensis* third stage larvae per day. This 'trickle infection' does not stimulate the immune system to reject the adults and the infection can be maintained for many weeks. Data from Jenkins, D.C. & Phillipson, R.F. (1970). *Parasitology,* **62**, 457–465.

tective measures, mainly associated with haemoglobin variations, occur. The best-known of these is the sickle-cell trait. Sickle-cell haemoglobin differs from normal haemoglobin in the substitution of a single amino acid, a valine for a glutamic acid, in the B chain, and this provides considerable protection against *Plasmodium falciparum*, the most pathogenic of the four species of malaria that affect man. In its homozygous form the sickle-cell gene is invariably fatal and, in the absence of malaria, natural selection rapidly eliminates it from the population. The additional selection pressure exerted by the parasite has meant that potentially fatal sickle-cell genes are maintained.

The complexity of the malaria life-cycle and the selective precision of the immune system complicate protective responses. Each stage produces its own antigens and there is considerable antigenic diversity and variation between species, strains and stages, so that immunity against one offers little protection against another but every stage is immunogenic and the host mounts a response to each. Invading sporozoites are surrounded by a circumsporozoite protein (CSP) which is highly immunogenic. The host response to it inhibits invasion of liver cells and thus prevents reinfection with the same strain of the parasite, another example of **concomitant immunity**, although it is often known as premunition when applied to malaria. The erythrocytic stages produce many stage specific antigens and it is the host's reactions to these, both B-cell and T-cell mediated, that prevent the infection becoming overwhelming and eventually bring it under control. Naturally acquired immunity, although incomplete, slowly develops and does much to protect the indigenous population.

Patients with chronic malaria infections often experience relapses even after removal from malarious regions. In some species, e.g. *P. vivax*, the relapses seem to be due to the re-awakening of latent schizonts in the liver which are stimulated into activity when immune suppression is relaxed following the acute phase of the disease. For other species including *P. falciparum* relapses appear to be due to antigenic variation in the small population of erythrocytic parasites which remain after immune elimination of the bulk of the infection. The host and parasite thus alternately gain the ascendancy with marked fluctuations in parasitaemia as the antigenic variant arises and is suppressed by the host's response.

As mentioned earlier, *Toxoplasma* may cause an acute, fatal infection in immunocompromised patients. These infections do not have to have been acquired after immune suppression but may be initiated from latent stages encysted in muscle and nervous tissue. In normal patients antibody responses to primary infections provide acquired immunity that may persist to prevent reinfection for life, but any immune depression may trigger the serious consequences of acute toxoplasmosis.

We have already seen that under some circumstances the host may present an effective immune response against gastro-intestinal nematodes, but such examples are relatively rare and there is little evidence for acquired protective immunity against adult intestinal platyhelminths. More commonly it is the tissue migration of larval stages that initiates immune responses but, before these can become fully effective, the adults escape to the relative safety of the gut lumen. Reinfection may thus result in damage and destruction of incoming larval stages while leaving the adults largely unscathed. There are advantages for the parasite in such a system as it prevents the host being over-burdened with parasites. If we accept that in most cases it is not in the parasite's interests to kill its host and thus destroy its own habitat, any mechanism that encourages this is likely to be selected for.

The filarial nematodes pose special problems both for their hosts and for parasite immunologists. Adults of species such as *Wuchereria* live in lymph

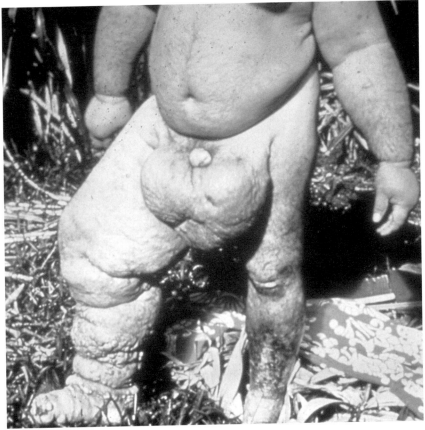

Fig. 6.9 Elephantiasis of the legs and scrotum as a consequence of infection with *Wuchereria bancrofti*.

nodes in the centre of the immune system, and one might expect them to be rapidly eliminated. In fact the reverse is true and they result in chronic infections of many years' duration. Our understanding of these infections is still far from complete as they are highly host specific and there is no truly appropriate laboratory model. Much of our knowledge has come from epidemiological studies of natural infections in endemic areas. Such studies have identified several groups of patients. A proportion are asymptomatic, have no circulating microfilariae (**amicrofilaraemic**), are serologically positive indicating exposure to infective stages, and appear to be naturally resistant. Others remain asymptomatic despite having circulating microfilariae, and the more gross deformities of **elephantiasis** (Fig. 6.9) are generally manifest in patients who no longer have circulating microfilariae. Circulating microfilariae thus do not correlate with overt disease symptoms. To complicate matters further, asymptomatic, microfilaraemic patients may present with another condition, **tropical pulmonary eosinophilia**, which can result in fibrosis of the lung and severe breathing difficulties if not treated with diethylcarbamazine, the drug often used against filarial infections.

Microfilariae are immunogenic and specific antibodies are produced against surface and sheath antigens, but these are often only seen after the patient has become amicrofilaraemic, possibly because in filaraemic infections they are adsorbed onto the large number of circulating microfilariae. *In vitro*, ADCC has been demonstrated against microfilariae but it is not clear how effective this is *in vivo* in man. Protective responses thus do occur but are very slow to develop. Why this is so is still uncertain and a number of possibilities have been proposed, for example, the microfilariae may evade recognition by changing their surface antigens or by disguising themselves by acquiring host molecules, and there is some evidence that both may occur. In addition infection may result in a general immunosuppression with consequent relaxation of many immune responses. Immunosuppression is not restricted to filarial infections and has been implicated in other chronic parasitic infections including malaria and trypanosomiasis.

Most ectoparasites spend too little time with an individual host for any protective immune responses to develop but all inject immunogenic salivary, often anticoagulatory, secretions. Hard ticks may feed for a week or so and this allows both T-cell and B-cell responses to develop to the antigens introduced by the tick. Much of the initial response that occurs around the site of the bite depends on infiltration of amine-rich basophils or mast cells and inflammatory responses mediated by IgE. The reaction is too slow to affect the overall feeding of the tick, which can engorge normally. However, other ticks subsequently feeding on the same host may not be so lucky: the host may be partially or completely immune to further attack and the tick may fail to engorge, may fail to moult normally or may die of desiccation, depending on the degree of immunity that has developed. Interestingly there is little stage specificity among tick antigens so that immunity induced by larvae is effective against nymphs or adults and *vice versa*.

The interactions between hosts and parasites are many and various; as hosts have evolved protective measures against parasite invaders the parasites have in turn developed counter measures to ensure their survival. There is continuous refinement on both sides and the mechanisms have become increasingly sophisticated and complex. When they are working efficiently they help to maintain a parasite population that can be safely tolerated by the host, but when they fail to work they can contribute to the pathological changes that are the subject of the next chapter.

7

Damaging effects of parasites

One of the main reasons that parasites have been so extensively studied compared with their free-living relatives is that they cause damage to their hosts, and when this results in disease to humans, their domesticated animals and their crops, action is called for. The effect of a single individual – with the exception of some ectoparasites, largely in their role as vectors, and a few large helminths – is rarely measurable. Parasitic protozoans reproduce in their hosts so that theoretically, a single invasive stage may result in damaging numbers. In practice such small infections are normally contained by the host's response and it is the intensity of infection that governs whether disease symptoms occur. Observable clinical damage for both protozoans and helminths is thus largely a parasite population phenomenon.

If one looks at the distribution of parasites in a population of hosts it quickly becomes apparent that they are not randomly distributed. Many hosts are not infected or have only a few parasites while a small proportion is heavily infected and carries a disproportionate parasite burden. This pattern, with the bulk of the parasite population aggregated into a small number of hosts, has now been recorded for various helminth species (Fig. 7.1) and appears to be generally applicable. Empirically the data closely fit a **negative binomial distribution** model in which the proportion infected (P) is dependent on the mean (m) and a measure of aggregation (k) such that $P=(1-(1+m/k)^{-k}$. The smaller the value of the more aggregated the population. Values of k from field studies have varied from about 0.1 to 1.0; the smaller values indicate that something like 80% of the parasite population is concentrated in only about 20% of the hosts. Inevitably it is these heavily infected individuals that are most likely to suffer overt disease.

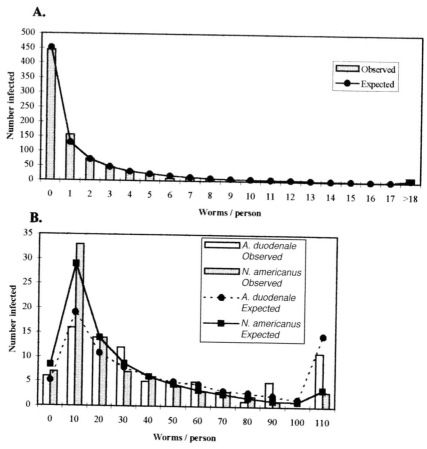

Fig. 7.1 Distribution of helminths within their host populations with fitted negative binomial distributions. A, *Ascaris lumbricoides* in man in Korea. B, Hookworm species in man in India. C, *Toxocara canis* in foxes. D, Microfilariae in mosquitoes. Data from: Seo, B. (1980); Schad & Anderson (as Fig. 7.2); Watkins, C. V. & Harvey, L. A. (1942) *Parasitology*, **34**, 155–79; Schmidt, W. D. & Robinson, E. J. (1972) *J. Parasitology*, **58**, 907–10.

Many factors, which are not easy to separate in individual cases, contribute to aggregation. In man social, environmental and behavioural factors may all affect exposure to invasive stages which themselves are likely to be discontinuously distributed in the environment, while genetic, nutritional and immunological factors affect susceptibility, parasite maturation and survival. Some individual hosts appear to be consistently more susceptible than others (Fig. 7.2); this persists even after treatment and has led to the concept of **predisposition**. In human communities there is evidence of multi-species pre-

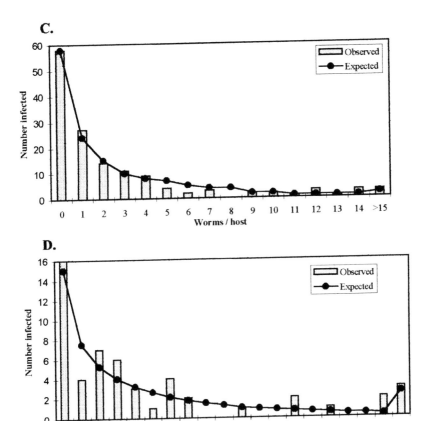

C.

D.

disposition suggesting that it is due more to a deficiency in the individual rather than to differential exposure. The present explanation for predisposition is thus based on variations in immunocompetence in dealing with the parasite internally rather than differences in infection levels.

A central tenet of parasitological dogma holds that it is not in a parasite's interests to fatally damage or significantly reduce the life-span of its host as, in doing so, it will destroy its own environment. A moment's reflection should lead to the conclusion that, like most dogmatic statements, this one is only partially true. Where the parasite relies on predator/prey transmission any factor that increases the chance of infected individuals being the prey victims serves to further the parasite's interests. Thus the space-occupying lesions in the brains of sheep caused by the coenurus of *Taenia multiceps* (Fig. 7.3) result in abnormal host behaviour, isolating the infected individual from

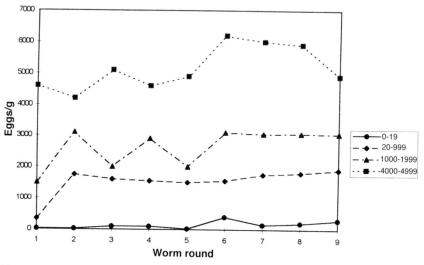

Fig. 7.2 Change in infection with time. Patients were assessed for hookworm infection during the first worm round and allocated to groups depending on the number of eggs found. These groups were maintained at each of the subsequent worm rounds over the next 18 months. There was no overlap between the groups; a lightly or heavily infected patient remained lightly or heavily infected throughout the study, supporting the idea that some individuals are more liable to infection than others. Data from Schad, G. A. and Anderson, R. H. (1985) *Science* **228**, 1537–40.

the flock and making it more liable to predation. Other examples were considered when we discussed transmission in Chapter 3.

A few parasites benefit directly from the death of their definitive host as we saw for *Trichinella spiralis* (see Chapter 3), but there are not many of them. One is *Capillaria hepatica*, a nematode that lives in the liver of rodent hosts. The adult females migrate through the liver parenchyma depositing eggs as they go. While some of the eggs may enter bile ducts and thus leave the host in the faeces, most have no route out of the liver and, in long-standing infections, the liver becomes full of egg-filled tracts. Transmission occurs when the host is eaten by a predator or scavenger. The eggs pass straight through unharmed, are voided in the faeces and are now infective for another rodent. There is little evidence that the parasite causes any significant damage. The mammalian liver accepts considerable redundancy and the life-span of most small mammals is insufficient for the parasite to cause enough damage to accelerate the host's demise, but the death of the host is required for transmission.

Most parasites do not lead to such extreme consequences, but the types of damage caused by parasites are many and various. A single species may cause a range of different effects depending on the intensity of the infection and

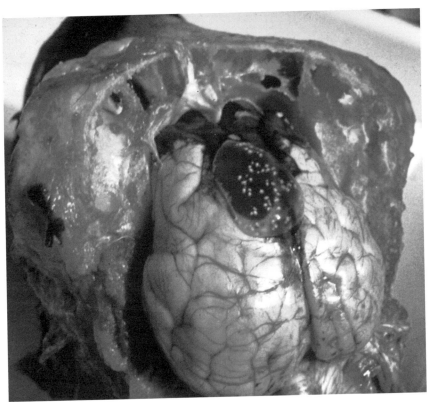

Fig. 7.3 *Coenurus cerebralis* in the brain of a sheep. This large cyst of the dog cestode *Taenia multiceps* causes behavioural changes in the intermediate host making it more liable to predation.

host specific factors such as genetic constitution, nutritional state and immunocompetence. For example, the major clinical consequence of chronic *Fasciola hepatica* in sheep and cattle is liver fibrosis. When metacercariae are ingested they excyst in the duodenum, cross the gut wall and enter the liver from the peritoneal cavity. The young flukes migrate through the liver *en route* to the bile ducts, destroying liver cells as they go. Once in the bile duct the spines on the tegument irritate the bile duct wall and the host responds by producing extra cells that increase the wall thickness and restrict further damage (Fig. 7.4). In cattle calcification of the bile duct often occurs, leading to the condition known as **pipe stem liver**. The loss of liver tissue means that liver function is disturbed, the infected animal fails to maintain blood plasma protein levels and there is often a build-up of fluid (**oedema**), especially under the jaw giving the condition known as **bottle jaw**. Adult flukes also cause

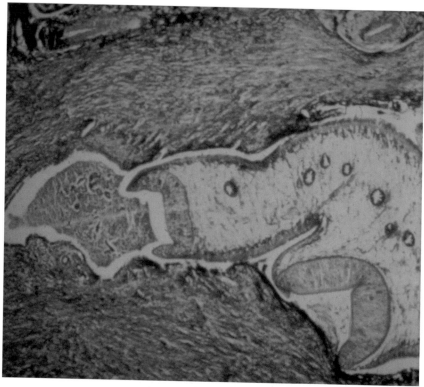

Fig. 7.4 Section through *Fasciola hepatica in situ* in the bile duct of its sheep host. The bile duct wall is much thickened as a consequence of the infection.

blood loss, about 0.5 ml/fluke/day, so that anaemia develops, especially if the host is on a less than optimal diet. It is the loss of condition, bottle jaw and the anaemia that give the outward signs of *Fasciola* infection, while the most serious effects are only manifest *post mortem*. If the host gets a massive dose of metacercariae in a short period of time the consequences may be catastrophic. The trauma caused by so many young flukes invading the liver at once results in such massive blood loss that the host may die suddenly with few outward signs of infection. The overall disease caused by *Fasciola*, fascioliasis, is thus a combination of a number of pathological changes that may manifest themselves in different ways in different individual hosts. The skill of the diagnostician comes in recognising the symptoms while there is still time to do something about the infection. *Post mortem* diagnosis does not help the patient!

For the sake of simplicity we can consider some of the pathological changes caused by parasites under a number of headings, but remember that these are selected examples. Most parasites cause more than one effect so that

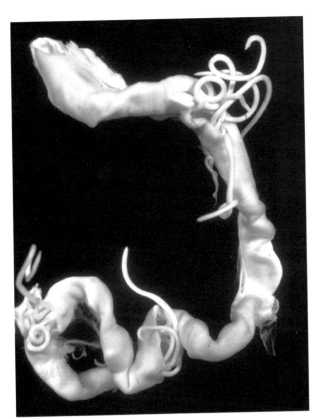

Fig. 7.5 Small intestine of a dog blocked by the presence of *Toxocara canis*.

the headings are not mutually exclusive and, just because a particular effect is not mentioned in any given example does not mean that it does not ever occur.

Physical and mechanical damage

The sheer mass of parasites may cause problems. Heavy infections of ascarid nematodes (Fig. 7.5) may sometimes form a compacted mass that can block the intestine which, in the absence of surgical intervention, may be fatal. Modern anthelmintics generally paralyse worms, but some of their predecessors increased worm activity and resulted in their literally tying themselves in knots, exacerbating the problem.

On a different scale, infection with *Giardia lamblia* (Fig. 5.3) may also be considered a problem of intestinal blockage. Many human infections are symptomless but in heavy infections virtually every cell in the duodenum is capped with an adherent protozoan so that normal absorption is impaired. Diarrhoea is a common symptom with copious, pale stools containing the fat that was not absorbed. Weight loss occurs, partly as a result of **anorexia** and partly from the malabsorption resulting from the infection.

There has been some argument about the causes of elephantiasis (Fig. 6.9), the extreme outward manifestation of infection with *Wuchereria bancrofti*, but blockage of the lymph nodes preventing normal drainage of the affected limb is certainly a contributory factor, although immunological reactions may also play a part. Continuous increased pressure in the lymph vessels eventually results in overgrowth of skin and subcutaneous tissues and leads to gross deformities, most commonly of the legs and scrotum, but the arms and the breasts in women may be affected.

One reason why *Plasmodium falciparum* is the most serious of the four species of *Plasmodium* that affect man is that it is responsible for **cerebral malaria**. The fine capillaries in the brain become blocked by masses of parasite-infected erythrocytes. The exact cause of the blockage remains somewhat controversial but schizont-infected erythrocytes develop small protrusions on their outer surface and these appear to prevent normal passage of the infected cells through the fine capillary beds. The infected cells then initiate a 'traffic jam', other cells build up behind them and eventually the whole capillary becomes inextricably blocked. The pressure that develops behind the jam may be sufficient to cause the capillary to burst and, if not treated rapidly, this may result in coma and death.

Even if there are insufficient parasites to cause blockage some may still cause mechanical damage; for instance, the proboscis of acanthocephalans may work its way right through the intestinal wall (Fig. 7.6), which may result in peritonitis and convert a mild infection into a potentially fatal one.

As mentioned earlier, one of the factors contributing to liver fibrosis in *Fasciola* infections is the continuous rasping of the bile duct wall by the spines in the tegument of the adult worms. In this case the irritation seems to offer little advantage to the parasite but in other examples advantages can be seen. The nocturnal migrations of *Enterobius vermicularis* females to the perianal skin to deposit their eggs cause irritation that results in the patient scratching the affected region and carrying the eggs back to the mouth. The tapeworm *Dipylidium caninum* utilises fleas and lice as intermediate hosts and relies on their ingestion for transmission to their canine hosts. The irritation caused by the movement and feeding of the ectoparasites encourages the dog to nip them which, while it is fatal for the ectoparasite, is crucial for the transmission of the cestode. A case of one parasite benefiting at the expense of another.

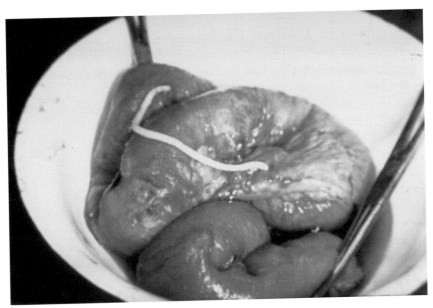

Fig. 7.6 Part of the small intestine of a man removed during surgery with an adult acanthocephalan, *Macracanthorhynchus hirudinaceus,* protruding through it. *M. hirudinaceus* is normally a parasite of the intestine of pigs but has recently been shown to be able to infect man. Human cases have been reported in areas in the Far East when extra dietary protein intake was encouraged by the ingestion of invertebrates, including beetles. Where these were infected with cystacanths and incompletely cooked transmission of the acanthocephalan became possible.

For some species irritation and its consequences are the main pathological features. The mange mites *Sarcoptes scabiei* of man and *Psoroptes ovis,* the caustative organism of sheep scab, both cause intense itching. Adult female *Sarcoptes* burrow superficially into the epidermis, feed on lymph and deposit eggs as they go. The larvae and nymphs which develop from the eggs continue the burrows until a series of tortuous tracts are formed. Early on the infection is largely symptomless but as the skin becomes sensitised to the presence of the mites a rash which causes an intense, continuous, almost unbearable itching develops, although not necessarily where the burrowing mites are. The scratching that the host uses to try to alleviate the effects often causes lesions which are liable to opportunistic secondary bacterial infection.

Psoroptes ovis live and feed on the surface of the skin of their hosts. They feed on lymph and induce intense local **inflammation** to which the host responds by rubbing and scratching against any available object. The inflamed lesion exudes serum which forms a crust, the notorious 'scab' after which the disease is named. The wool becomes loosened and falls, or is pulled out, following the attentions of the sheep. So much time is spent concentrating on relieving the irritation that feeding may almost cease; weight gains are thus markedly affected and there have been cases where infected animals have died of starvation because they were unable to compete for limited winter feed.

Anaemia

The blood feeding habits of many parasites leave their hosts liable to the effects of anaemia. A single adult *Necator americanus* causes the loss of about 0.05 ml of blood per day while each *Ancylostoma duodenale*, the other common hookworm of man, consumes as much as 0.2 ml per day. The worms are profligate feeders and much of the blood that is ingested passes straight through them into the host's intestine where up to half of the iron may be reabsorbed. However, the continuous drain on iron reserves, especially in areas of the world where the diet lacks adequate nutritional iron, eventually results in clinical anaemia. Pregnancy and menstruation obviously exacerbate the problem. The patient shows all the classic signs of developing anaemia: tiredness, aching muscles, breathlessness and possible oedema and, as the condition worsens, pallor, especially of the mucous membranes. The normal level of haemoglobin is about 16 g/100 ml for men and 14 g/100 ml for women. In severe hookworm anaemia this may be reduced to as little as 2 g/100ml.

Blood feeding ectoparasites rarely result in clinical anaemia although populations of poultry mites (*Dermanyssus gallinae*) may become so large in unhygienic poultry houses that young birds in particular may be liable to **exsanguination**. Ticks also have the potential to cause considerable blood loss. Each adult tick completing its life-cycle will remove some 1–3 ml of blood from its various hosts. Infestations of several thousand ticks on a single horse or cow are not unheard of and fatalities, especially during times of nutritional stress, have been known. Natural infections are generally much smaller than this and the main significance of ticks is in their ability to transmit other organisms, of which more later.

There are other causes of parasite-induced anaemia. When a *Plasmodium* schizont bursts to release its merozoites it destroys the erythrocyte that housed it. Uninfected red cells in the vicinity may also be lysed and a reduction in the number of new red cells produced in the bone marrow also contributes to a developing anaemia which is especially serious in children. Interference with red cell production is also at the root of the pernicious anaemia induced in some patients by the presence of the pseudophyllidean cestode *Diphyllobothrium latum* (see Chapter 5).

Immunological damage

On several occasions mention has been made of pathology following sensitisation of the host by the presence of the parasite. Recent immunological studies have demonstrated that an over-enthusiastic or inappropriate immune response may be detrimental to the host and **immunopathology**, a new branch of disease study, has been established. There is insufficient space here to deal with more than a single example, but immune damage is becoming much more widely recognised in parasite infections.

By way of example let us consider a pair of *Schistosoma mansoni* recently established in the mesenteric vein of a naïve host. During invasion they have shed larval tegumental membranes and alerted the host to their presence, but have incorporated host molecules into their adult tegument so that they are largely immunologically invisible. The adult female rests in the gynecophoric canal of the male and extends into the finest venules to lay her eggs. The eggs pass easily through the thin venule walls and, aided by the spines on their egg shells and released enzymes, work their way to the gut lumen. The host responds to the presence of the eggs: a granuloma consisting of motile macrophages, plasma cells and eosinophils forms around them and helps to carry them to the lumen. Once there the granuloma is discarded and the eggs pass out in the faeces. The host mounts a humoral response to the large amounts of schistosome antigen and this phase of the infection is marked by rather non-specific symptoms including fever and chills, fatigue, **eosinophilia** and diarrhoea.

As the infection continues passage across the gut wall becomes increasingly difficult, and the host responds to the continuous bombardment of egg antigens by laying down fibrous tissue. This results in fewer eggs completing the journey to the lumen and more being swept away in the blood stream to be filtered out in the capillary beds of other organs, especially the liver. Granulomata form around the eggs trapped in the liver (Fig. 7.7). T helper cells react to the antigens secreted by the miracidia and released through pores in the egg shell. The cytokines produced form a focus for lymphocytes, eosinophils and macrophages. Eventually the miracidium in the egg at the centre of the granuloma is killed so that granuloma formation can be considered protective, but the damage done to the liver may be considerable, with many of the hepatocytes replaced by graulomatous tissue. The immune

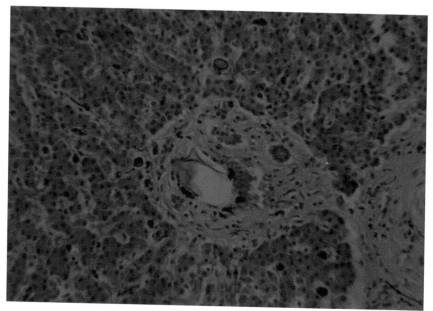

Fig. 7.7 Granulomatous tissue (light in colour) around a spined *Schistosoma mansoni* egg in the liver of a man. Much of the pathology associated with schistosomiasis in man is due to granulomatous reactions to eggs in the tissues.

response has additional effects: tumour necrosis factor released by the macrophages seems to stimulate egg production in the females but is inimical to invading schistosomules. The adult parasites, bathed as they are in what should be a highly hostile medium, actually exploit the immune response that should be acting against them to maintain their own healthy population, without fear of overcrowding, and to enhance dissemination of their progeny. Damage is almost entirely the result of host responses to the eggs.

Physiological disturbances

It can be argued that anaemia and most other forms of parasite disease are manifestations of physiological damage. However, there are some parasites that cause specific physiological changes and it is these which will now be addressed. For example, infection with *Ascaris lumbricoides* causes a wide range of intestinal disturbances resulting in decreased absorption of protein, fat, carbohydrate and vitamin A, all of which may lead to slow growth in children as *A. suum* infections do in pigs. Additionally and specifically *Ascaris* directly interferes with lactose digestion. Infected patients exhibit an intolerance for

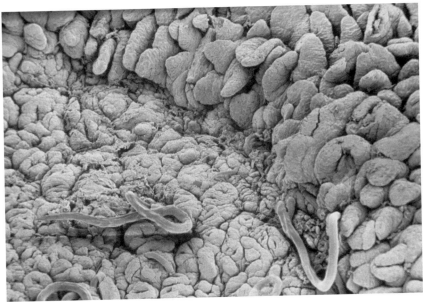

Fig. 7.8 Villous atrophy. Region of the small intestine of a sheep infected with the nematode *Trichostrongylus vitrinus* showing normal villi to the top right and flattened shortened villi in the vicinity of the worms. Photograph by the Moredun Institute, Edinburgh, reproduced with permission.

lactose, and experience diarrhoea, gas and acute intestinal discomfort after consuming milk or milk products. The condition can be relieved by anthelmintic treatment to remove the worms, although it takes some time for symptoms to disappear completely. This obviously has a considerable effect on children, for whom milk is often a considerable component of the diet.

In uninfected hosts water and ions are absorbed as they pass through the small intestine, so that by the end of the ileum the osmolarity of the fluid in the gut lumen tends towards that of blood plasma. To maintain this condition water may pass both ways across the gut wall. Any factor that upsets this balance is likely to interfere with normal intestinal function and often manifests itself as diarrhoea. To overwhelm the absorptive ability of the large intestine and result in fluid stools requires at least five times the normal volume of liquid to be delivered to it. The blockage of the duodenal wall by *Giardia*, mucosal epithelial erosion caused by coccidia, **villous atrophy** (Fig. 7.8) caused by *Trichostrongylus vitrinus* and absorptive changes following *Ascaris* infection all result in diarrhoea as one symptom.

The abdominal discomfort resulting from the presence of intestinal parasites may lead to a loss of appetite and inappetance or anorexia has been reported for a large number of host/parasite combinations. In some cases this is mild and self-limiting but in others it may result in the complete cessation of feeding. The precise mechanisms that result in anorexia in parasitised animals are not fully understood. In healthy hosts the control of appetite is centred in two areas in the hypothalamus, the region at the base of the brain that is central to much of the control of hormonal balance in the body. One of these areas, known as the feeding centre, initiates feeding, while the other, the satiety centre, ends it. The two centres do not work independently but are linked by a series of complex hormonal feedback mechanisms. The relative importance of hormonal imbalance and pain caused by the presence of parasites in initiating anorexia in parasitised hosts is still not fully resolved.

While hormonal changes are only suspected of contributing to cases of anorexia they have been proven to have far-reaching effects in other host/parasite associations. *Lymnaea stagnalis*, the molluscan intermediate host for the bird schistosome *Trichobilharzia ocellata*, fails to produce eggs when infected. Control of maturation of the female reproductive organs in uninfected snails is effected by a hormone secreted by the dorsal bodies of the cerebral ganglia, while ovulation and oviposition are regulated by a second hormone produced from caudodorsal cells. A third hormone, calfluxin, controls the albumin gland, a secondary sex organ that produces nutriment for the developing embryo. Infected snails have increased dorsal body activity but the effect of the hormone on the reproductive system is reduced, while both the caudodorsal hormone and calfluxin levels are reduced. At present it is not apparent whether the parasite itself produces an agent(s) that competes with normal host receptor sites and thus inhibits the dorsal gland hormone effects or whether it induces the host to produce such an agent.

The plerocercoids of *Ligula intestinalis* in the body cavity of fish intermediate hosts (Fig. 3.12) also result in parasitic castration. There is suppression of gonadal development and gonadotrophin-producing cells in the pituitary of both male and female fish. Again the precise mechanism is uncertain. Attempts to isolate a parasite-produced sex steroid that could suppress gonadotrophin production have proved fruitless and, while a number of other possible mechanisms have been suggested, conclusive evidence is still lacking.

Possibly the most dramatic of all of the effects of parasite-induced hormonal changes are seen in mice infected with the plerocercoids of *Spirometra mansonoides* (Fig. 7.9). These impressive animals are the consequence of a growth factor, plerocercoid growth factor, that ensures that the mice continue to grow excessively. Plerocercoid growth factor has been extensively studied and both chemically and genetically it is very closely related to human growth hormone. So close is the relationship that it has been suggested that at some

Fig. 7.9 Mouse carrying 11 plerocercoids of *Spirometra mansonoides*. The parasites release a growth factor that results in excessive body growth. Photograph from Mueller, J.F. (1963). *Annals of the New York Academy of Sciences,***113**, 217–233.

time in its evolutionary past *Spirometra* has incorporated the human growth hormone gene into its own genome. If this is true then *Spirometra* must have been one of the pioneer genetic engineers.

Tissue invasion and cell destruction

Four-fifths of patients infected with *Entamoeba histolytica* show no symptoms, the amoebae live as harmless commensals in the large intestine feeding on bacteria on the mucosal surface. The faeces of such patients contain cysts that are infective to others. As a result of factors that are still not fully understood

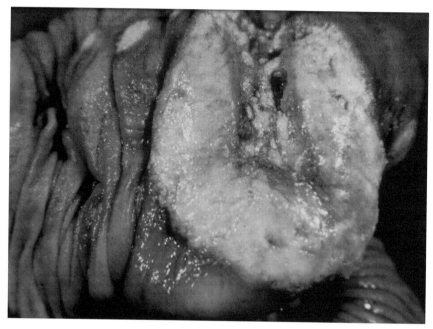

Fig. 7.10 Ulceration of the intestine resulting from *Entamoeba histolytica* infection. The deep ulcers may provide access to the circulatory system when the parasites may be carried to other organs, especially the liver, and initiate abscesses.

a carrier may suddenly develop acute disease. The benign commensals produce powerful proteolytic enzymes, become invasive and produce characteristically flask-shaped ulcers (Fig. 7.10). The trophozoites change from feeding on bacteria to ingesting red blood cells that abound in the ulcers and may get carried in the blood stream around the body. Typically they pass to the liver where they erode abscesses, but similar lesions have been reported in many other organs. Ironically these invasive forms do not produce cysts, although the copious dysenteric stools are full of trophozoites that do not survive for long outside the host, and in any case cannot pass the stomach even if they are ingested. It is thus an unsuspecting carrier who is responsible for the transmission of the disease. The one exception to this occurs amongst homosexual men when direct transmission on the genitalia can convert amoebiasis into a sexually transmitted disease.

Mention has already been made of the destructive effects of *Plasmodium* schizonts on red blood cells. Coccidians cause similar problems to the intestinal cells that they inhabit. The extent of the damage depends on the precise cells that are parasitised. For example in the chicken all of the parasitic stages of *Eimeria praecox* are spent in superficial epithelial cells and, while they may increase cell turnover and cause some intestinal dysfunction, they are rarely truly pathogenic. *Eimeria tenella* on the other hand invades epithelial cells deep in the crypts of the caeca and when cells containing the second generation schizonts and the gametocytes burst they cause very considerable cellular destruction, haemorrhage and not infrequently the death of especially young hosts.

After ingestion by their sheep and cattle hosts third stage larvae of *Ostertagia* spp. invade the abomasal gastric glands. During the summer months they emerge from the glands about three weeks after infection but during the autumn and winter they may spend several months in a hypobiotic state in the abomasal wall. The presence of the larvae in the glands results in specialised secretory cells being replaced by undifferentiated mucus-producing cells and a massive thickening of the abomasal wall. These morphological changes are accompanied by biochemical changes. The destruction of the glands means that little pepsinogen or HCl are produced. The pH of the abomasal contents rises markedly as there is so little HCl. This means that any pepsinogen that is produced is not converted to pepsin. Not unsurprisingly both anorexia and diarrhoea accompany the infection. The watery diarrhoea is often bright green as there is insufficient acid in the abomasum to denature the chlorophyll in the feed.

Disease transmission by arthropods

While parasitic arthropods may cause direct damage their greater significance is as vectors of other pathogens. Table 7.1 summarises some of the organisms involved. Hard ticks transmit more organisms than any other arthropods. Most of these are animal pathogens although some, like *Babesia* and *Borrelia burgdorferi,* are also transmissible to man. Human infections with *Babesia* have only occurred in patients whose spleen has been surgically removed, normally following traumatic injury, and there is no evidence of the protozoan being infective to intact persons. *Borrelia burgdorferi*, the causative organism of **Lyme Disease**, has come to prominence in recent years, especially in the USA although it has also been reported from Europe and Africa. It was first recognised officially in the USA in 1975 in children in Lyme, Connecticut, and it was not until 1982 that the causative organism was identified. Since its recognition as a distinct disease a wave of cases has been reported in Canada and 43 states across America wherever its deer tick intermediate host occurs. There has been an extensive public education programme to alert the public to the possible danger.

Table 7.1. *Organisms transmitted by arthropods*

Arthropod	Organisms transmitted						Disease
	Viruses	Spirochaetes	Rickettsia	Bacteria	Protozoa	Helminths	
Crustacea							
Copepoda						*Dracunculus* *Diphyllobothrium* *Ligula* *Spirometra*	Guinea worm Broad (fish) tapeworm
Decapoda						*Paragonimus*	Lung fluke
Arachnida							
Argasidae (soft ticks)		*Borrelia recurrentis*					Relapsing fever
Ixodidae (hard ticks)	*Flavivirus*	*Borrelia burgdorferi*	*Rickettsia rickettsii* *Coxiella burneti*	*Francisella tularensis*	*Babesia* *Theileria*		Louping ill Lyme disease Spotted fever Q fever Tularaemia Red water fever East Coast fever
Mites			*Rickettsia tsutsugamushi*				Scrub typhus

Vector	Virus	*Borrelia*	*Rickettsia* / *Bartonella*	*Yersinia*	Protozoa	Helminth	Disease
Insecta Anoplura (lice)		*Borrelia recurrentis*	*Rickettsia prowazeki* *Rickettsia quintana*				Epidemic relapsing fever Epidemic typhus Trench fever
Hemiptera (Reduviidae or assassin bugs)					*Trypanosoma cruzi*		Chagas' disease
Diptera *Anopheles* (mosquitoes)					*Plasmodium*	*Wuchereria* *Brugia*	Malaria Filariasis
Culicines (mosquitoes)	Yellow fever Dengue Encephalitis						Yellow fever Dengue
						Wuchereria *Brugia* *Onchocerca* *Loa loa*	Filariasis
Simulium (black flies) *Chrysops* (horse flies) *Phlebotomus* (sand flies)			*Bartonella bacilliformis*		*Leishmania* *Trypanosoma*		River blindness Loiasis Bartonellosis
Glossina (Tsetse flies) Siphonaptera (fleas)	Myxoma		*Rickettsia typhi*	*Yersinia pestis*		*Dipylidium caninum*	Leishmaniasis African trypanosomiasis Myxomatosis Endemic typhus
							Plague

As vectors of human diseases the mosquitoes probably reign supreme today. They are responsible for the transmission of a greater range of human pathogens than any other group of arthropods. However, fleas, as transmitters of plague, and lice, the vectors of typhus, have both had an enormous effect on human affairs, the latter especially in times of war and deprivation, and have helped to change the course of history.

The consequences of parasite damage, however caused, are such that action needs to be taken to reduce or eliminate them. We will examine some of the approaches and pitfalls of parasite contol in the final chapter.

8

Problems of parasite control

Many of us who live in the developed world will go through our entire lives without knowingly contracting a parasitic infection. Those, such as toxoplasmosis, cryptosporidiosis and toxocariasis that we do encounter usually produce transitory, non-specific symptoms that, in the absence of complications, are likely to be diagnosed as something more common. The nuisance value of mosquitoes and midges may be more obvious, as anyone who has tried camping in the highlands of Scotland on a calm August night will testify, but they are not life threatening. If, on a foreign holiday, we are unfortunate or foolhardy enough to become infected with malaria, schistosomiasis or any of the many other tropical parasitic diseases we can generally obtain chemotherapy on our return home. This should leave us wiser but little the worse for wear. On the other hand there are millions of people in the so-called Third World who do not know what it is like to live without multiple parasitic companions. Many of these infections are at sub-clinical levels and are well tolerated by their hosts, but they provide a reservoir of infective stages for new and more vulnerable patients.

Most domesticated animals also carry a range of parasites. Some of these may be transmissible to man, and these are **zoonoses** and of direct human concern, but most threaten the health of animals or crops and thus indirectly the well-being of their owners.

While many parasitologists find parasites intrinsically fascinating and the study of both parasites and their relationships with their hosts rewarding, even the most enthusiastic have to accept that the objects of their passion are not held in universal esteem. Most of the general public together with many members of the medical and veterinary professions hold with the adage that 'the only good parasite is a dead parasite'.

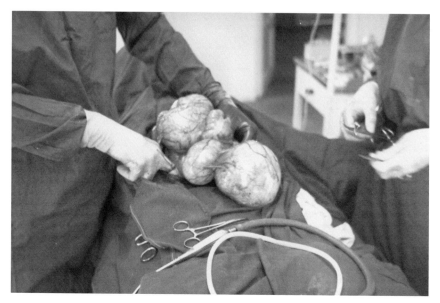

Fig. 8.1 Human hydatid cysts exposed prior to surgical removal. Photograph by Professor C. Macpherson, reproduced with permission.

Approaches to parasite control

At an individual level when a patient, animal or human, presents with the symptoms of a parasitic infection, treatment is nearly always chemical. A few conditions may be treatable by physical, surgical means, e.g. removal of the nodules that form around *Onchocerca volvulus* in the skin, removal of adult *Loa loa* as they migrate across the eye beneath the conjunctiva and removal of hydatid cysts (Fig. 8.1), but usually drugs are prescribed. Drugs, however, all have their drawbacks. While many are highly effective they are expensive, invariably have side-effects even if these are mild, and are contraindicated for some groups of patients; many need expert administration and, if used inappropriately, especially in large-scale control programmes, may result in additional problems of resistance.

It is relatively easy to define what we mean by control if considering an individual. In humans and companion animals we expect the results of our control measures to result in the cessation of symptoms and the eradication of all parasites from the patient. Control at a population rather than an individual level may have different priorities and as the size of the host population increases from local to global then the questions become more difficult to answer.

The first question that needs to be considered is what constitutes acceptable control? Eradication is the obvious ideal. The Guinea worm, *Dracunculus medinensis*, is now an endangered species and few, if any, conservationists have raised a whisper of protest. In 1986 there were about 3.5 million cases in the World. In 1991 a programme was initiated by the United Nations World Health Organization (WHO) with the stated intention of eradicating dracunculiasis, as infection with *Dracunculus* is called, by the end of 1995.

Transmission of *Dracunculus* involves contamination of drinking water with larvae that have escaped from adult female worms that protrude through blisters in the skin of their hosts. The usual method of contamination comes when people wade into still water during their daily lives. The larvae are ingested by copepods, especially *Cyclops*, and develop to third stage larvae that are infective when the *Cyclops* is ingested in drinking water. Man is the only common host for *Dracunculus*. A number of instances of other animals, including dogs, being infected with the same or similar parasites have been reported but there is little evidence that these contribute to human infection. The life-cycle is relatively short, being complete in about a year. Eradication is thus theoretically a simple matter. All one needs to do is ensure that no infected person contaminates drinking water supplies. While that is easy to say it is extremely difficult to implement. *Dracunculus* had a vast if patchy distribution across tropical Africa, the Middle East, India and Pakistan. By 1990 it was largely eliminated from the Middle East and the numbers in India and Pakistan had been dramatically reduced, but that still left much of Africa between the equator and the Tropic of Cancer. National active case searches have been undertaken in endemic countries and intervention measures introduced. There is no very effective chemical, anthelmintic treatment for dracunculiasis and the traditional remedy of winding the large adult worm slowly out of the lesion on a stick is fraught with difficulties if undertaken on a large scale, so most attention has been concentrated on the provision of safe drinking water. Deep wells and pumped water help to prevent contact between patients and drinking water supplies, but field workers are still liable to drink from small ponds if these are convenient. Education programmes have been introduced to try to persuade people to change their habits and to introduce them to the idea of boiling or filtering drinking water. Boiling is rarely feasible as fuel is usually scarce but filtering is a practical way of removing the infected intermediate hosts and is being widely encouraged and adopted. The filters do not need to be sophisticated – an old T-shirt will do. *Cyclops* can also be removed from water by chemical treatment and this approach is also being used where other measures have not been successful. The provision of safe drinking water, of course, also has extra benefits and prevents the transmission of many other infectious organisms. The overall campaign is enormous and expensive and involves national governments, international aid organisations and multinational companies but progress has been steady and, while the end of 1995 deadline was not met, the latest WHO figures suggest that on 31st December 1995 there were about 100,000 cases left, a reduction of some 97% on the 1986 figure. The days of *Dracunculus* do seem to be numbered.

Not every eradication campaign has had such an encouraging outcome. In the 1950s and 60s, following successful pilot trials in Europe and Central America, the WHO initiated and supported an attempt at global malaria eradication. The programme, which was embraced by more than 60 malarious nations, was mainly directed against the vector, *Anopheles* mosquitoes, using the long-lasting insecticide DDT with a back-up using the cheap and effective drug chloroquine against the parasite itself. After some 15 years following enormous expenditure of time, effort and money the programme ended in failure and was officially scaled down from an eradication campaign to a control programme. The main reason for the failure was that the complexity of the problem, especially the biology of the vectors, was underestimated. For example, insecticide spraying was concentrated in and around houses. The idea was that when a female *Anopheles* was replete from her blood meal she would rest on the insecticide-treated walls, contaminate herself and die before the parasite had a chance to complete its transmission cycle. Unfortunately not only did mosquito strains evolve that were resistant to DDT but behavioural differences also emerged where female mosquitoes flew into houses, fed and flew straight out again without touching the insecticide. In addition, species that never entered houses proved to be much more efficient malaria vectors than had been anticipated. Control of these would have required blanket insecticide spraying which was impracticable but was also ruled out on environmental grounds. Concerns about the dangers of residual insecticides were being forcibly expressed at the same time. To add to the woes of the eradicators, under the selection pressure exerted by over-usage and under-dosage *Plasmodium falciparum*, the most pathogenic of the four human malaria species, developed resistance to chloroquine. There was no alternative treatment available so that mass chemotherapy was not a viable alternative control strategy. In 1972 the Global Eradication of Malaria Programme was officially declared dead and since then individual nations have made their own arrangements to protect, or not, their citizens.

The limited host range, relatively short life-cycle and easily identifiable transmission sites of *Dracunculus* allowed eradication to be considered feasible, but considerable scepticism had to be overcome when the programme

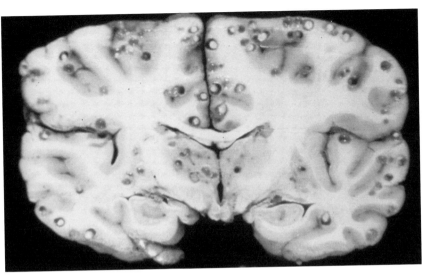

Fig. 8.2 Human neurocysticercosis caused by cysts of the cestode *Taenia solium* in a human brain. Cysticercosis is one of only two parasitic infections considered by a recent working party to be potentially eradicable given money and international goodwill.

was first proposed. Such conditions cannot apply to other parasites. An international task force recently concluded that of all the human parasitic diseases considered only filariasis (Fig. 6.9) and taeniasis/cysticercosis (Fig. 8.2) are potentially eradicable given national goodwill and international co-operation. For all other parasites less rigorous but more realistic criteria for control have to be accepted. Most control programmes limit their aspirations to controlling the damage caused by the parasite rather than eliminating it completely. For example, in farm animals, while the removal of overt symptoms is paramount, it may be acceptable to the farmer for a small population of parasites to remain as long as they do not affect his production costs. This so-called economic control has financial advantages but may be dangerous as it has led to short cuts being taken in both formulation and frequency of drug administration, with the consequent establishment of ideal conditions for resistance to develop.

Resistance to antiparasitic drugs has been increasing at an alarming rate. So far, with the exception of antimalarial drugs mentioned above, the impact on human chemotherapy has been relatively slight as the drugs concerned have not been widely misapplied, but concern is increasingly being expressed in veterinary circles where resistance, especially to antinematode drugs, is rapidly becoming a world-wide problem.

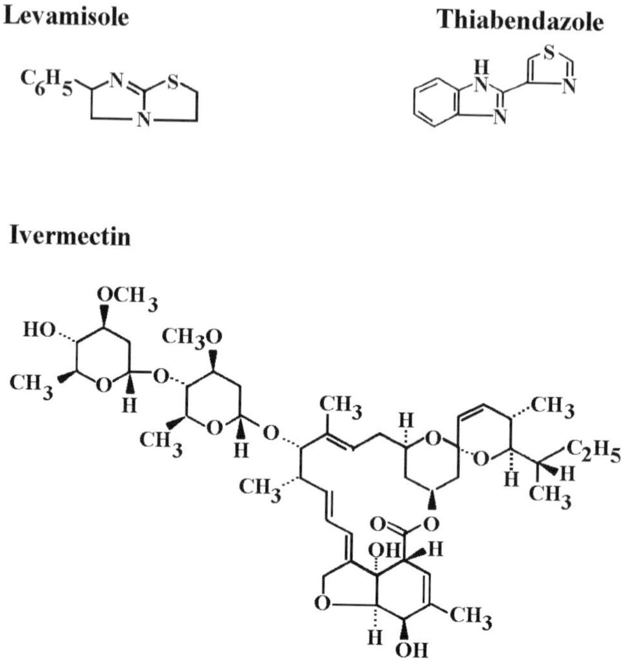

Levamisole

C_6H_5—N—S—N

Thiabendazole

Ivermectin

Fig. 8.3 Chemical structure of examples of the three main groups of broad spectrum anthelmintics. Levamisole is an imidazothiazole, thiabendazole a benzimidazole and ivermectin an avermectin.

There are three main families of broad spectrum anthelmintic drugs used against veterinary nematodes, the **benzimidazoles**, the **imidazothiazoles** and the **avermectins** (Fig. 8.3). Despite very different chemical structures and modes of action, resistant strains have arisen against all of them and multi-resistant strains are now emerging. With widespread use resistance can develop very rapidly: for example, although ivermectin was only introduced in 1981 resistance began to appear in South Africa in 1986 and since then it has been reported in South America, Australasia, the USA and Europe. It is thought that the gene or genes that confer resistance are present in the parasite population before they ever meet the drug and that resistant strains are selected from individuals carrying these genes. Initially the frequency of resistant individuals is low but under continuous exposure to the same drug the selection pressure increases the fitness of heterozygous resistant individuals which, on further exposure give rise to homozygotes and the population becomes increasingly resistant. Probably the single most important feature in selecting resistant strains is misguided under-dosing. This may result from mistakes in formulation or administration, attempts at economising or from misconceptions about drug efficacy in different species. For example, there has been a tendency to treat goats with similar drug levels to sheep, but differences in the metabolism of goats mean that reduced amounts of drug are available in the intestinal lumen of goats. The overall result is under-dosing, which probably explains why there is such a high prevalence of anthelmintic resistance in goat herds. Indeed, in some areas in New Zealand, goat rearing has had to be abandoned as there is no effective anthelmintic left to control nematode damage. Relatively few large-scale surveys have been conducted in the UK but, while the situation appears to be less dire, there is still reason for concern. Benzimidazole resistance has been recorded in about 66% of goat herds surveyed, in about 50% of sheep flocks in southern England and 25% of Scottish flocks, and in increasing numbers of horses. Although rarer, **ivermectin** and multiple benzimidazole/ivermectin resistance are also beginning to appear.

Part of the reason that drug resistance is causing such concern is that new antiparasitic drugs are not appearing on the market. The prohibitive cost of developing and testing new drugs for human use has meant that many multinational pharmaceutical companies have cut back on research and development of products that are likely to be mainly used in the less developed world, which is least equipped to pay for them. It is no coincidence that ivermectin was released for veterinary use long before it was approved for human treatment. This does not mean that there are no new antiparasitic drugs on the horizon. One which has caused some excitement recently is artemisinin, a highly effective antimalarial drug extracted from a Chinese medicinal plant, qinghao (*Artemisia annua*). Initial clinical trials with artemisinin and some of its derivatives have been encouraging and have proved especially useful against multi-resistant *P. falciparum*. The results of further trials are needed for universal approval but its proponents are optimistic.

Drug companies continually strive to produce pharmacologically active compounds and, with increasing knowledge of the molecular structure of parasites and their essential components, it is becoming possible to use computer graphics to help to design drugs with highly specific activities. Even these have to go through the lengthy, rigorous and exceedingly expensive trials required before they can be approved for either animal or human use.

Alternative methods of control

If eradication is impracticable and chemotherapy is becoming less attractive due to cost and drug resistance, is there any hope for the 1000 million or so humans and countless domesticated animals that suffer from parasitic infections? The answer rests with prevention, breaking the life-cycle and halting transmission. That, of course, was the concept behind the ill-fated malaria eradication campaign and is much easier said than done. There has been a reawakening of interest in relatively low technology, sustainable control measures for both human and veterinary parasites which will reduce reliance on chemotherapy and thus hopefully extend the useful life of the drugs that are available. Filtering water to remove *Cyclops* and prevent *Dracunculus* transmission is one such example as is the use of mosquito nets, with or without insecticide treatment, to reduce *Anopheles* bites and malaria.

In many parts of the world parasite transmission is seasonal. Free-living stages and vectors are susceptible to climatic conditions so that dry or cold seasons may halt transmission for part of each year. In temperate regions especially not only is transmission seasonal but the extent of disease varies from year to year. The factors governing the seasonality of the diseases caused by *Fasciola hepatica* and *Nematodirus battus* in the UK are sufficiently well known and predictable to allow forecasting of disease outbreaks.

In the case of fascioliasis forecasts are based mainly on the rainfall in early summer, specifically June. This has been shown to be the single most important factor that controls the population of the intermediate snail host, *Lymnaea truncatula*, and it is the snail population that in turn governs the extent of *Fasciola* infection. Most disease symptoms are seen in the late autumn and winter (Fig. 8.4), caused by developing adults which were ingested during late summer and early autumn as metacercariae that had emerged as cercariae from snails infected by miracidia during the summer. A forecast can be prepared from data collected in June that predicts disease six to nine months later, plenty of time for avoiding action to be taken. Some infections may overwinter in the snails and these emerge as cercariae when the average daily temperature reaches 10°C in the spring. These encyst as metacercariae on the pasture during the early summer and give rise to a minor peak in disease symptoms in late summer/early autumn. The significance of this disease peak can also be predicted by consideration of autumn and spring rainfall statistics.

For *Nematodirus battus* the crucial factor in the UK is the temperature in March. *N. battus* eggs only hatch when the mean daily temperature rises to 10°C following a period of chilling below 0°C. Once the temperature reaches the 10°C threshold there is a rapid rise in egg hatch, and a sharp peak of infective larvae on the pasture (Fig. 3.4). If this peak coincides with the new season's lambs being on the pasture there is likely to be severe nematodiriasis. In a warm, early spring the larval peak occurs while many lambs are still suckling and not likely to ingest large numbers of larvae, but in a late spring the

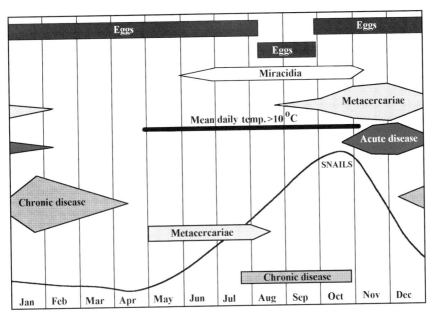

Fig. 8.4 Epidemiology of fascioliasis in sheep in the UK. The population of parasites is determined by the population of snails which in turn is dependent on the wetness of the pasture, especially in the early summer. There is no development below 10°C but overwintering may occur in the snail.

lambs will be grazing extensively and disease incidence will be high. An estimate of the lateness of the spring can be achieved by measuring the March earth temperature at two distant points across the country and calculating a disease index from the recorded results. The data can be further refined to additionally give an estimate of the date when peak hatch will occur. Prediction does not control parasites but it does give the farmer a chance to instigate control measures at appropriate times and thus limit the blanket use of anthelmintics.

The mathematical formulae used to predict fascioliasis and nematodiriasis are simple and were derived empirically from observations of disease outbreaks and climatic conditions. Much more sophisticated models have been developed for other diseases, especially those of man, and these provide additional support to any control programme.

Cultural techniques including rotation of pasture, ploughing, and replanting, and improving pasture by drainage, removing a hay or silage crop or alternately grazing sheep and cattle when the parasite species are not shared, may all be employed to reduce infestation. None will eliminate the parasites but may be sufficient to keep them within commercially acceptable limits.

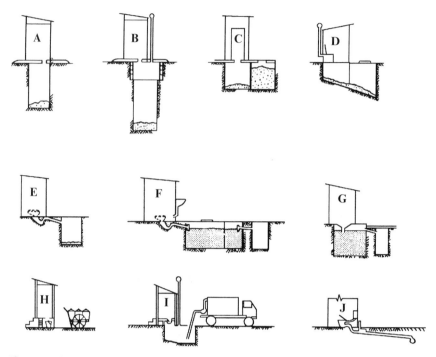

Fig. 8.5 Various sanitation systems. Upper row, dry on-site systems: A, pit latrine. B, VIP latrine. C, twin pit latrine. D, compost toilet. Centre row, wet on-site systems: E, pour flush toilet. F, septic tank and soakaway. G, aqua privy. Lower row, off-site systems: H, bucket latrine. I, vault toilet. J, sewerage. Redrawn from Cairncross, S. (1987). *Parasitology Today*, **3** (3), 94–98.

Such procedures are hardly applicable to human diseases. There is little doubt that hygienic waste disposal and the provision of safe drinking water would have a greater effect on human health than any other single measure. Many bacterial and viral diseases, as well as parasites such as *Entamoeba*, *Ascaris*, *Trichuris* and hookworm with direct life-cycles and many platyhelminths with indirect life-cycles, would all be controllable. It is not necessary to universally install the infrastructure of a full sewerage system. There are many less sophisticated systems which, if conscientiously installed and maintained, can provide perfectly adequate and hygienic alternatives (Fig. 8.5) and are much more appropriate in many areas where water is at a premium. Even the notorious pit latrine can be made more acceptable by some simple modifications. The **VIP latrine** is not just for visiting dignitaries (the name comes from Ventilated, Improved Pit latrine). This design overcomes two of the main problems with the basic pit latrine, smell and flies. A ventilation pipe creates a down draught that removes smells and a fly-proof mesh across its top prevents mass fly entry. The progeny of any flies that do enter and breed fly towards light but, if the latrine itself is suitably dark, they fly towards the light at the top of the vent pipe where they are trapped by the underside of the mesh and eventually die.

Provision of facilities is one thing, obtaining acceptance and universal use is quite another. Their purpose is totally undermined if they are less acceptable to the local population than the bush alternative. No system will remain hygienic if not maintained and an education campaign and follow-up programme are essential if the most is to be obtained from the initiative.

Most large-scale control campaigns against human parasites have been aimed at reducing transmission by removing vectors or intermediate hosts. Since the scaling down of the Global Malaria Eradication campaign the largest single international scheme has been the West African *Onchocerca* Control Programme (OCP). This began in 1974 in seven countries in West Africa (Benin, Burkina Faso, Côte d'Ivoire, Ghana, Mali, Niger and Togo). It was extended in 1986 to include Guinea, Guinea-Bissau, Senegal and Sierra Leone and is due to run for 30 years, beyond the length of life of any adult *Onchocerca* alive at the start. Initially the programme concentrated on the black fly vector, *Simulium*, and consisted of weekly applications by ground and air of an ecologically acceptable larvicide, **temephos** (**Abate**), to water courses in which the larvae live. Resistance to temephos was first reported in 1980 but alternatives, including the highly specific bacterial insecticide ***Bacillus thuringiensis*** serotype H.14, were available and careful manipulation and rotation have ensured that vector control remains efficient and the main arm of the programme. In 1987 ivermectin was approved for human use and was donated by its manufacturers Merck, Sharp and Dohme for *Onchocerca* control. Ivermectin is a very efficient killer of microfilariae with minimal side-effects and an annual oral dose is sufficient in most areas to prevent eye lesions and reduce skin microfilariae. This not only prevents transmission but also reduces the risks of onchocercal skin lesions, which may be uncomfortable and disfiguring.

The programme is expensive, currently some US$30 million annually, but the return on the investment has been calculated as some 20%, so that in hard economic terms it is successful. In more meaningful human terms it is considered that up to 200,000 people have been saved from blindness, over 30 million people are now protected from ocular lesions and 1.5 million who were initially infected are no longer so. Twenty-five million hectares of riverine valleys which were uninhabitable because of the flies and onchocerciasis have been made available for settlement and the disease has been eliminated as a public health problem from much of the originally treated area. The success of the OCP has been such that other endemic countries in Africa have been looking enviously on and, with the co-operation of WHO, are now considering their own programmes based on ivermectin treatment rather than *Simulium* control. How successful these will be remains to be seen.

While vector and intermediate host control have received most attention, definitive host control has been employed in attempts to control zoonoses. Large game culls have been tried in abortive attempts to reduce *Trypanosoma* transmission, and more successfully dog control has been used to control hydatid disease. Man is infected with hydatid cysts from ingestion of *Echinococcus granulosus* eggs from dog faeces. In the early 1970s the island of Cyprus had the dubious distinction of having one of the highest human hydatid infection rates in the world. The main definitive hosts were large free-ranging dogs and most of the older sheep had hydatid cysts. The livers and lungs of home-slaughtered sheep were regularly fed to dogs so that the animal cycle had ideal conditions for perpetuation and man inevitably became incidentally infected. The main thrust of the Cyprus hydatid control programme was directed against the dogs and was enforced by law. In the first 18 months 42,870 dogs (93 per cent of the total) were put down and all other dogs had to be leashed or enclosed at all times when they were not working. They were inspected several times a year and any infected with *Echinococcus* killed. Slaughtering of sheep was centralised into licensed, supervised village abattoirs and infected offal incinerated. By the use of a multifaceted education programme and rigorous application of the law *Echinococcus* has been eliminated from Cyprus. Such draconian measures are not essential and in New Zealand, another country that has made concerted efforts to reduce hydatid disease, a worming programme for dogs is in force (Fig. 8.6). This programme has not had such dramatic success but considerable progress in hydatid reduction has been achieved, and there are now indications that after 35 years *Echinococcus* has been eliminated from New Zealand.

Immunological control

The great hope for the future of parasite control lies with vaccine development, but as was pointed out in Chapter 6, parasites have been living with their hosts for a long time and have evolved many ways of evading the effects of immune responses against them.

A live radiation-attenuated vaccine, **Dictol**, against *Dictyocaulus viviparus*, has been both scientifically and commercially successful. A similar vaccine against

Fig. 8.6 Dogs awaiting anthelmintic treatment against *Echinococcus granulosus* in New Zealand. Photograph by Dr M.A. Gemmell, reproduced with permission.

the dog hookworm, *Ancylostoma caninum*, was developed in Glasgow but manufactured, released and marketed in the USA. While it dramatically reduced adult hookworm levels it did not produce sterile infections and proved commercially non-viable when vets failed to accept vaccination, preferring to manage hookworm infection with anthelmintic treatment. The vaccine was discontinued after about two years.

A commercial live vaccine has recently become available for use against the protozoan *Toxoplasma* in sheep. The original parasite was isolated from the foetal membranes of an aborted lamb. After more than 3000 passages in mice it was found to have lost the ability to develop normally and establish permanent infections but did still trigger a full protective immune response. It is now grown in tissue culture, administered by injection and provides protection for at least 18 months. Toxoplasmosis is, of course, a zoonosis, causing problems especially for pregnant women. Live vaccines are not acceptable for human use but the success of veterinary vaccines has provided encouragement for the search for synthetic vaccines. Again development has been faster in the veterinary than the medical field due largely to the less rigorous criteria that need to be satisfied before release for animal use.

Taenia ovis is a cestode from the intestine of dogs. It gained its specific name from the intermediate stage, *Cysticercus ovis*, which develops in the heart, diaphragm and other muscles of sheep. The adults have little effect on dogs but the presence of the cysts results in the rejection of the meat and thus has financial implications for livestock farmers in intensive sheep rearing areas. The life-cycle of *T. ovis* is similar to that of other *Taenia* species, with eggs releasing oncospheres in the sheep intermediate host. These invade the tissues and develop into cysts that are ingested by dogs and mature in the intestine. Crude extracts of *T. ovis* oncospheres have been used successfully to immunise sheep against subsequent challenge with infective eggs. Such a natural vaccine requires a continuous, plentiful supply of eggs to produce the oncospheres, and about ten years ago a programme was initiated in Australia to produce the necessary quantities of antigen using the new recombinant DNA technology.

Careful fractionation of the original oncospheral extracts demonstrated that effective antigens were to be found amongst polypeptides with a relative molecular mass (M_r) in the range 47–52 kDa. Hatched and activated oncospheres were then used as a source of mRNA from which a library of cDNA clones was established. These clones were screened using purified rabbit anti-*T. ovis* antibody probes and two of the expressed antigens, GST-45W and GST-45S were isolated and tested in an immunological trial in sheep. Of these two GST-45W was the more effective and gave about 70% protection against challenge infections compared with controls. Further trials using different dose rates and different methods of administration increased the efficacy to >90%. The early isolates had low yields and tended to be unstable but further refinement of the cDNA recombinant antigen and modified extraction techniques have provided acceptable yields of a vaccine, now known as GST-45B/X, that retains its potency (>90% protection) for at least 18 months at 2–8°C. A similar recombinant vaccine has been reported to give 95–96% protection against *Echinococcus granulosus* hydatid cysts in sheep. This may have implications for the control of hydatidosis in man in due course.

An undoubted scientific success, the *T. ovis* vaccine, rather like the live attenuated hookworm vaccine mentioned above, has not as yet been a commercial success. Again the problem is more related to other vested interests and alternative control strategies than to the efficacy of the product.

Similar logical and scientific processes have been applied to the development of vaccines against other parasites. For example an integral membrane glycoprotein of M_r 110,000 has been extracted from the sheep abomasal nematode *Haemonchus contortus*. Known as H11, this protein, when injected into sheep, has been shown to confer immunity against larval challenge. H11 has been cloned and trials are in hand. These recombinant vaccines are an exciting development but release of one for human use is still some way away.

An alternative approach to synthetic vaccine production has been used by a group working on malaria in Colombia. Using chemical methods they analysed sporozoite and merozoite proteins and then synthesised a peptide polymer containing amino acid sequences from both. This vaccine, SPf66, has been shown to be safe and immunogenic in both *Aotus* monkeys and humans, but how effective it is against *Plasmodium falciparum* under field conditions is still a subject of some controversy. Trials in South and Central America and in an area of intense malaria transmission in Tanzania have suggested that up to 30% protection may be gained by the use of SPf66, but in the Gambia, under different transmission conditions, no protection was achieved in children under one year of age. Other large-scale trials are in progress and the jury is still out.

While the Brave New World of recombinant and synthetic vaccines is still awaited, parasite control must depend on more traditional methods. One of the more powerful weapons in the armoury of the controllers is the education of the population into the biology and significance of parasites and the need for control measures to be implemented and maintained. In its own way, that is what this book has been trying to achieve.

Further reading

The time taken in writing and producing a text book means that it is out of date before it is even published. The only way to keep up to date with research ideas is to read the current literature in specialised journals, and a selection is listed below. Some of these are devoted exclusively to parasitology while others often contain parasitological papers. Some, e.g. *Advances in Parasitology, Parasitology Today* and the *Symposia of the British Society for Parasitology*, are essentially review publications and are very useful sources of up to date information on selected topics. Journals inevitably assume a knowledge of the language of the subject and this may be more easily acquired from the books listed. Some of these are general texts while others address more restricted topics.

Journals containing papers, abstracts and reviews on parasites

Acta Tropica
Advances in Parasitology
American Journal of Tropical Medicine and Hygiene
Annals of Tropical Medicine and Parasitology
Biological Abstracts
Bulletin of the World Health Organization
Canadian Journal of Zoology
Experimental and Applied Acarology
Experimental Parasitology

Fundamental and Applied Nematology
Helminthological Abstracts
International Journal for Parasitology
Journal of Comparative Biochemistry and Physiology
Journal of Helminthology
Journal of Infectious Diseases
Journal of Nematology (mainly plant parasitic nematodes)
Journal of Parasitology
Medical and Veterinary Entomology
Molecular and Biochemical Parasitology
Nematologica (mainly plant parasitic nematodes)
Parasite
Parasite Immunology
Parasitology
Parasitology Today
Proceedings of the Helminthological Society of Washington
Protozoological Abstracts
Symposia of the British Society for Parasitology
Systematic Parasitology
Transactions of the Royal Society of Tropical Medicine and Hygiene
Tropical Diseases Bulletin
Veterinary Parasitology
Veterinary Research
Zeitschrift für Parasitenkunde: Parasitology Research

General comprehensive textbooks

Cox, F. E. G. (1993). *Modern Parasitology.* Blackwell Scientific Publications, Oxford.
Melhorn, H. (1988). *Parasitology in Focus.* Springer-Verlag, Berlin.
Peters, W. & Gillies, H. M. (1995). *A Colour Atlas of Tropical Medicine and Parasitology.* Mosby-Wolfe, London.
Roberts, L. S. & Janovy, J. (1996). *Foundations of Parasitology.* Wm. C. Brown, Dubuque, IA.
Smyth, J. D. (1994). *Introduction to Parasitology.* Cambridge University Press, Cambridge.
Soulsby, E. J. L. (1982). *Helminths, Arthropods and Protozoa of Domesticated Animals.* Bailliere-Tindall, London.

Trager, W. (1988). *Living Together. The Biology of Animal Parasitism.* Plenum Press, New York.

Urquhart, G. M., Armour, J., Duncan, A. M. & Jennings, F. W. (1987). *Veterinary Parasitology.* Longman, Harlow.

Specific topics

Anderson, R. M. & May, R. M. (1992). *Infectious Diseases of Humans, Dynamics and Control.* Oxford University Press, Oxford.

Barnard, C. J. & Behnke, J. M. (1990). *Parasitism and Host Behaviour.* Taylor & Francis, London.

Behnke, J. M. (1990). *Parasites: Immunity and Pathology.* Taylor & Francis, London.

Bryant, C. & Behm, C. (1989). *Biochemical Adaptation in Parasites.* Taylor & Francis, London.

Esch, G., Bush, A. & Ho, J. (1990). *Parasite Communities: Patterns and Processes.* Chapman & Hall, London.

Hyde, J. E. (1990). *Molecular Parasitology.* Open University Press, Milton Keynes.

Kennedy, C. R. (1976). *Ecological Aspects of Parasitology.* North Holland, Amsterdam.

McAdam, K. P. W. J. (1989). *New Strategies in Parasitology.* Churchill Livingstone, Edinburgh.

Marr, J. J. & Muller, M. (1995). *Biochemistry and Molecular Biology of Parasites.* Academic Press, London.

Pike, A. W. & Lewis, J. W. (1994). *Parasitic Diseases of Fish.* Samara Publishing, Tresaith, Dyfed.

Scott, M. E. & Smith, G. (1994). *Parasitic and Infectious Diseases: Epidemiology and Ecology.* Academic Press, San Diego.

Taylor, A. E. R. & Baker, J. R. (1987). *In vitro Methods for Parasite Cultivation.* Academic Press, London.

Toft, C. A., Aeschlimann, A. & Bolis, L. (1991). *Parasite–Host Associations. Coexistence or Conflict?* Oxford University Press, Oxford.

Wakelin, D. (1996). *Immunity to Parasites.* Cambridge University Press, Cambridge.

Warren, K. S. (1993). *Immunology and Molecular Biology of Parasitic Infections.* Blackwell Scientific Publications, Oxford.

Individual groups of parasites

Protozoa

Coombs, G. & North, M. (1991). *Biochemical Protozoology.* Taylor & Francis, London.

Gillies, H. M. & Warrell, D. A. (1993). *Essential Malariology.* Edward Arnold, London.

Hide, G., Mottram, J. C., Coombs, G. H. & Holmes, P. H. (1997). *Trypanosomiasis and Leishmaniasis: Biology and Control.* CAB International, Wallingford, Oxford.

Kreier, J. P. (1977–1995). *Parasitic Protozoa. Vols 1–10.* Academic Press, New York, London.

Smith, D. F. & Parsons, M. (1996). *Molecular Biology of Parasitic Protozoa.* Oxford University Press, Oxford.

Thompson, R. C. A. & Reynoldson, J. A. (1993). *Giardia: from Molecules to Disease.* CAB International, Wallingford, Oxford.

Helminths

Platyhelminthes

Arme, C. & Pappas, P. W. (1983). *Biology of the Eucestoda. Vols 1–2.* Academic Press, London.

Jordan, P., Webbe, G. & Sturrock, R. F. (1993). *Human Schistosomiasis.* CAB International, Wallingford, Oxford.

Smyth, J. D. & Halton, D. W. (1983). *The Physiology of Trematodes.* Cambridge University Press, Cambridge.

Smyth, J. D. & McManus, D. P. (1989). *The Physiology and Biochemistry of Cestodes.* Cambridge University Press, Cambridge.

Thompson, R. C. A. & Lymbery, A. J. (1995). *Echinococcus and Hydatid Disease.* CAB International, Wallingford, Oxford.

Nematoda

Crompton, D. W. T., Nesheim, M. C. & Pawlowski, Z. S. (1989). *Ascariasis and its Prevention and Control.* Taylor & Francis, London.

Kennedy, M. W. (1991). *Parasitic Nematodes: Antigens, Membranes and Genes.* Taylor & Francis, London.

Lee, D. L. & Atkinson, H. J. (1976). *Physiology of Nematodes.* MacMillan, London.

Lewis, J. W. & Maisels, R. M. (1993). *Toxocara and Toxocariasis.* Institute of Biology, London.

Schad, G. A. & Warren, K. S. (1990). *Hookworm Disease: Current Status and New Directions.* Taylor & Francis, London.

Acanthocephala

Crompton, D. W. T. & Nickol, B. B. (1985). *Biology of the Acanthocephala.* Cambridge University Press, Cambridge.

Parasitic Arthropoda

Burgess, N. R. H. & Cowan, G. O. (1993). *A Colour Atlas of Medical Entomology.* Chapman & Hall Medical, London.

Kettle, D. S. (1995). *Medical and Veterinary Entomology.* CAB International, Wallingford, Oxford.

Lehane, M. J. (1996). *Biology of Blood Sucking Insects.* Chapman & Hall, London.

Service, M. W. (1996). *Medical Entomology for Students.* Chapman & Hall, London.

Walker, A. (1994). *The Arthropods of Humans and Domestic Animals.* Chapman & Hall, London.

The Internet contains a considerable amount of information about parasites. A useful introductory home page which has links to other pages has been established by the British Society for Parasitology. The BSP page can be found at http://www.umds.ac.uk/bsp.

Glossary

Abate   Insecticide used against *Simulium* larvae in the Onchocerciasis Control Programme. Also known as temephos.

Acanthella   Intermediate stage in the life-cycle of an acanthocephalan between the acanthor and the cystacanth.

Acanthocephala   A phylum of parasitic worms with characteristic anterior spiny proboscis.

Acanthor   Stage in an acanthocephalan life-cycle that hatches from the egg.

Acetabulum   Alternative name for the ventral sucker of a digenean.

Amicrofilaraemic   Without microfilariae circulating in the blood.

Aneurysm   Blood filled sac produced by dilation of the weakened wall of a blood vessel.

Anorexia   Lack of appetite.

Antibody   Immunoglobulin produced by a B-cell that binds to a specific antigen.

Antibody Dependent Cell Mediated Cytotoxicity (ADCC)   Cytotoxic reaction in which an antibody-coated target cell is killed by an Fc-bearing leukocyte.

Antigen   A molecule that is recognised by a T-cell receptor or an antibody.

Antigen Presenting Cell (APC)   A cell that presents processed antigens and MHC molecules to T-cell receptors.

Apical complex   Organelles possessed by apicomplexan sporozoites used to penetrate cell walls.

Apicomplexa   Alternative name for protozoans of the Sporozoa Group.

Arrested development   The halting of development of some nematodes to overcome adverse environmental conditions.

Aspidogastrea   One of the minor classes of the phylum Platyhelminthes

Avermectin   One of the families of broad spectrum anthelmintic drugs.

Bacillus thuringiensis   Bacterial insecticide.

Basophil Granulocyte found in the blood which, like a mast cell, may be triggered to release its biologically active granules when stimulated by IgE and antigen.

Benzimidazole One of the families of broad spectrum anthelmintic drugs.

Black head Disease in turkeys caused by infection with *Histomonas meleagridis*.

Bothrium (plural bothria) Dorsal or ventral sucking groove on the scolex of pseudophyllidean cestodes.

Bottle jaw Swelling of the space below the lower jaw as a result of *Fasciola* infection.

Brood capsules Structure in a hydatid cyst in which the individual scolices develop and mature.

Bursa of Fabricius The primary lymphoid organ in birds, the site of B-cell maturation.

Cement glands Glands in male acanthocephalans that secrete a sealing cement to block the female opening after mating.

Cephalothorax Fused head and thorax found in chelicerate arthropods.

Cercaria Second free-living digenean larval stage produced by asexual reproduction in a sporocyst or redia in the mollusc intermediate host.

Cerebral malaria Serious and often fatal complication in *Plasmodium falciparum* infections in which the small capillaries in the brain become blocked by infected red cells.

Cestoda The tapeworms: a major class of the phylum Platyhelminthes.

Chalimus larva Specialised crustacean larva adapted for attachment to its host by an anterior filament.

Chelicerae Pincer-like appendages on the anterior limbs of chelicerate arthropods.

Chemotherapy Treatment of a disease using chemicals.

Cilia Fine threads linked by intracellular organelles that act as locomotor organs for ciliated protozoans and to transport particles in metazoans.

Circadian migration Regular daily migration, specifically applied to the migrations of *Hymenolepis diminuta* in the rat gut.

Cloaca Combined reproductive and intestinal opening found in male nematodes.

Coccidia A large family of mainly intestinal sporozoan protozoans with a direct lifecycle and oocyst transmission stages.

Coenurus Hollow bladder-like cestode cystic stage in which several scolices bud from an internal germinative layer.

Colony Stimulating Factor (CSF) Factor that allows the proliferation and differentiation of haematopoietic cells.

Commensalism Inter-specific association in which both partners are able to lead independent lives but one or both may benefit when in association with the other.

Concomitant immunity Resistance to infection aimed at invading organisms while allowing an existing infection to be maintained.

Contaminative transmission Direct transmission following contamination of food, water or fingers with faecal material.

Coprophilic Attracted to faecal material.

Coprophobic Having an aversion to faecal material.

Copulatory bursa Cuticular expansion at the posterior of some male nematodes, used to orientate the female during mating.

Coracidium Ciliated oncosphere hatching from the egg of pseudophyllidean cestodes.

Cuticle Outer multi-layered membrane of the nematode body wall. Lost at each moult.

Cystacanth Stage in acanthocephalan life-cycle infective for the definitive host.

Cysticercus Hollow bladder-like cestode cystic stage with a single invaginated scolex.

Cytokine Low molecular weight proteins that stimulate or inhibit the immune system.

Dauer larva A quiescent stage entered into by some nematode larvae while enclosed in the cast cuticle of the previous stage.

Definitive host In most cases the host in which a parasite reaches sexual maturity. In some cases the definitive host is the one of most significance to man.

Dermis Vascularised layer of the skin beneath the epidermis.

Dictol Commercial live vaccine for use against the nematode *Dictyocaulus viviparus*.

Digenea The class of Platyhelminthes that includes the flukes, with indirect life-cycles and usually mollusc intermediate hosts.

Dioecious Having separate male and female sexes.

Diploid Having chromosomes in pairs.

Direct life-cycle Life-cycle with a single host.

Ectoparasite A parasite that lives on the outer surface of its host.

Elephantiasis Massive size increase in parts of the body, commonly the legs and scrotum, due to chronic infection with filarial nematodes.

Embryophore Part of a cestode egg shell wall, often thick and protective.

Endoparasite A parasite that lives within the body of its host.

Eosinophilia Elevated level of eosinophils in the blood, often associated with chronic parasite infection.

Epidemiology Study of the distribution, prevalence and transmission of diseases.

Epidermis The outermost unvascularised region of the skin consisting of several cellular layers bounded by the stratum corneum.

Epitope Part of an antigen that is recognised by an antigen receptor.

Erythrocytic schizogony Asexual division of malaria parasites within their host's blood stream.

Excretory pore Pore on the ventral surface of a nematode through which the excretory system opens.

Exocytosis Discharge from a cell of particles too large to diffuse through the cell wall.

Exsanguination Extensive loss of blood.

Faeco-oral transmission Direct contaminative transmission with faecal material ingested.

Felt-fibre layer Central fibrous layer in the tegument of an acanthocephalan.

Flagellum (plural flagella) Long fine thread that by lashing activity acts as a locomotor organ for flagellate protozoans.

Flukes Digenean parasites.

Gametogony Development of sporozoan merozoites into male and female gametocytes.

Germinal layer Inner layer of cestode cysts that bud off scolices.

Glycocalyx Filamentous carbohydrate layer found on the outer surface of many cells.

Gynecophoric canal The groove formed by the folding round of the body margins of
a male schistosome in which the female resides.

Halteres Small club-shaped balancing organs that replace the second pair of wing in
dipterous flies.

Haploid Having a single set of unpaired chromosomes.

Helminth Parasitic worm, generally restricted to members of the Platyhelminthes,
Nematoda and Acanthocephala.

Hemimetabolous Gradual metamorphosis in arthropods with immature forms
appearing similar to the adults.

Hermaphrodite Possessing both male and female sexual organs.

Hexacanth Six-hooked oncosphere hatching from a cestode egg.

Histamine Vasoactive amine produced and released by basophils and mast cells.

Holometabolous Complete metamorphosis during insect development.

Host specificity Range of hosts in which a parasite can develop.

Host stream Plume of odour produced down wind of an animal.

Hydatid cyst Cystic stage of the cestode *Echinococcus*. May grow to enormous propor-
tions and produce large numbers of individual protoscolices.

Hydatid sand Individual protoscolices produced in a hydatid cyst.

Hydrostatic skeleton Skeletal structural rigidity produced by the incompressible nature
of fluids.

Hypobiosis State of reduced metabolic activity attained to prolong life, often over a
period of environmental stress.

Hypodermis The layer in the body wall of a nematode that secretes the cuticle.

Hypostome One of the mouthparts of an acarine, toothed in ticks, not in mites.

Imidazothiazole One of the families of broad spectrum anthelmintic drugs.

Immediate hypersensitivity (IH) Rapidly produced immune over-reaction that may result
in undesirable consequences including, in extreme cases, anaphylactic shock and
death.

Immunogen Any substance that stimulates an immune response.

Immunoglobulin A protein produced by B-cells that acts as an antibody.

Immunopathology The study of interactions between the immune system and disease.

Indirect life-cycle Life-cycle with two or more essential hosts.

Inflammation Tissue response to injury or damage to tissue. The classical signs are
pain, redness, heat and swelling.

Inquilism An inter-specific association in which one animal lives within the shell or
body of another: synoecious commensalism.

Interferon (IFN) One of a series of cytokines with anti-viral properties: IFNγ also
acts to regulate the immune system.

Interleukin Cytokine secreted by leukocytes.

Intermediate host Host in which development but not sexual maturity occurs.

Inter-specific associations Associations between animals of different species.

Intra-specific associations Associations between animals of the same species.

Intrinsic factor A component of gastric secretions required for the absorption of vitamin B_{12}.

Keds Blood sucking, often wingless, dipterous flies belonging to the family Hippoboscidae.

Lacunar canals Canal system in the tegumant of acanthocephalans, probably acting as a circulatory system.

Lemnisci Paired bodies extending from the anterior of acanthocephalans whose function is still problematical.

Leukocyte A white blood cell.

Lice Small dorso-ventrally flattened ectoparasitic insects belonging to the orders Mallophaga and Anoplura.

Ligament sacs Connective tissue tubes in which the gonads of acanthocephalans develop.

Lyme disease Tick-transmitted disease caused by *Borrelia burgdorferi*.

Lymphocyte White blood cell produced in the bone marrow and processed to produce either B-cells that mature into plasma cells and release immunoglobulins or T-cells that stimulate or inhibit the immune system through the release of cytokines.

Macrogamete Large, passive 'female' gamete resulting from a macrogametocyte.

Macrogametocyte Cell giving rise to a macrogamete.

Macrophage Phagocytic blood cell which may also act as an antigen presenting cell.

Major Histocompatibility Complex (MHC) A genetic region that codes for molecules responsible for cell recognition and antigen presentation.

Mast cell Tissue cell containing numerous granules that may be discharged when stimulated by IgE and antigen.

Membrane Attack Complex (MAC) Complex of complement components that perforate the cell membrane of target cells and result in lysis.

Memory cell Cloned T- or B-cell produced during a primary infection that remains to mount a rapid secondary response on exposure to the same antigen.

Merozoite Daughter cell produced by schizogony.

Metacercaria Infective stage for the definitive host in a digenean life-cycle, often quiescent and encysted.

Metacestode Cystic stage that develops from the oncosphere in the life-cycle of a cestode.

Microfilaria (plural microfilariae) First larval stage of a filarial nematode, normally found in the blood or tissues of the definitive host.

Microgamete Active 'male' gamete produced from a microgametocyte.

Microgametocyte Cell giving rise to a microgamete.

Micropredator Temporary ectoparasite.

Microthrix (plural microthriches) Minute projections on the tegument of a cestode that greatly increase the surface area.

Miracidium Motile, ciliated stage that hatches from a digenean egg, infective for the molluscan intermediate host.

Monogamous Having only a single mate.

Monogenea Class of Platyhelminthes with direct life-cycles, largely parasites of fish.

Monophyletic Originating from a single phylum.

Muscularis mucosa Muscle layer bordering the gut mucosa.

Mutualism Inter-specific association, obligatory on both sides, in which both partners benefit.

Myiasis Infection by fly larvae.

Neck The unsegmented region behind the scolex of cestodes where the proglottides develop.

Negative binomial distribution Frequency distribution used to describe over-dispersed aggregated populations in which the variance is greater than the mean.

Nematoda Ubiquitous phylum of unsegmented pseudocoelomate worms.

Oedema Accumulation of fluid in extracellular spaces in the body.

Oesophageal glands Glands in the oesophageal tissue of nematodes whose secretions may aid digestion or host invasion.

Oesophagus Alternative name for the pharynx of nematodes

Oncomiracidium Ciliated larva of mongenean platyhelminths.

Oncosphere Cestode larval stage that emerges from the egg.

Ontogenetic migration Migration during development, specifically applied to the anterior migration of *Hymenolepis diminuta* during development in the rat gut.

Oocyst The cyst that results from sexual reproduction in sporozoan protozoans.

Oocyte Cell which after meiosis forms the egg.

Oogenesis The production of eggs.

Ookinete The motile zygote of *Plasmodium*.

Operculum Cap on the egg shell of some parasites through which the larva emerges.

Opisthaptor Posterior attachment organ of monogenean platyhelminths.

Opisthosoma Posterior region of the body of an acarine.

Oral infection Transmission of a parasite through the mouth.

Oral sucker The suckers that surround the mouth of a digenean.

Ovarian fragments The individual remnants of the embryonic ovary that circulate in the body cavity of female acanthocephalans.

Ovijector Muscular structure in the uterus of some female nematodes that sorts the eggs before expelling them through the vulva.

Parasitism Inter-specific association in which one partner benefits at the other's expense.

Parasitoid Organism parasitic during early life that kills its host later, frequently an insect parasitic in another insect.

Paratenic host Inessential host in which a parasite can survive but does not develop, also known as a transport host.

Parthenogenetic Ability to reproduce without necessity of fertilisation of the egg.

Pathogenicity Capacity of an organism for producing disease.

Pedipalps   Second pair of appendages on the head of arachnids.

Percutaneous infection   Transmission of parasites through the skin.

Periparturient rise   Increase in nematode egg production seen during late pregnancy and about the time of birth in sheep and cattle. Sometimes called spring rise.

Peritonitis   Inflammation of the peritoneum, the membrane lining the abdominal cavity.

Pernicious anaemia   Anaemia resulting from the inability of the body to absorb adequate vitamin B_{12} resulting in large numbers of large immature red blood cells in the circulation.

Phagocytosis   Ingestion of particles by cells by invagination of the cell membrane.

Pharynx   Anterior muscular pump used in the ingestion of food by some helminths. In nematodes also known as the oesophagus.

Pheromone   Substance secreted by one animal that has a physiological effect on another of the same species.

Phoresis   Inter-specific association based on transportation.

Photophobic   Having an aversion to light.

Photophilic   Functioning best in light.

Pinocytosis   The uptake of extracellular fluid by the formation of invaginations in the cell membrane.

Pipe-stem liver   Thickened and calcified bile duct due to the presence of a parasite.

Plasma cell   Differentiated B-cell that secretes large amounts of antibody.

Platyhelminthes   The phylum of normally dorso-ventrally flattened helminths that contains the Monogenea, Digenea and Cestoda amongst others.

Plerocercoid   The larval stage of a pseudophyllidean cestode that develops from the procercoid and is infective for the definitive host.

Polyphyletic   Originating from more than one phylum.

Predilection site   The site usually occupied by a parasite on or in its normal host.

Predisposition   The tendency for an individual to be susceptible to a particular infection.

Pre-erythrocytic schizogony   Asexual reproduction that occurs in some sporozoan protozoans prior to invading blood cells. In *Plasmodium* it occurs in the liver.

Proboscis receptacle   Sac in the anterior body of an acanthocephalan into which the proboscis may be retracted.

Procercoid   Larval stage of a pseudophyllidean cestode that develops in the first intermediate host from the coracidium.

Proctodeal feeding   Process whereby a drop of fluid is passed from the hind gut of one termite to another so that mutualist protozoans may be transferred.

Proglottid (plural proglottides)   Individual segment of a cestode containing a set of hermaphrodite reproductive organs.

Prosoma   Anterior body region of an arachnid consisting of the cephalothorax and bearing the appendages.

Protista   Kingdom of single-celled organisms.

Protoscolex (plural protoscolices)   Juvenile scolex that buds off within a cestode cyst.

Pseudocoelom   Fluid filled body cavity of a nematode.

Pseudopodium (plural pseudopodia) Temporary protrusion in a cell wall associated with flowing movements of protoplasm and locomotion in amoebae and phagocytic cells.

Purine Twin ring chemical compound with the empirical formula $C_5H_4N_4$ which, in substituted forms, produces the essential purine bases adenine and guanine that are required for the production of nucleic acids.

Pyloric sphincter The sphincter muscle that closes the distal aperture of the stomach at its junction with the duodenum.

Pyrimidine Single hexagonal ring compound with the empirical formula $C_4H_4N_2$ which, in substituted forms, gives the essential bases uracil, cytosine and thymidine which are required for nucleic acid production.

Radial layer Thick inner layer of the acanthocephalan tegument that contains the lacunar canals and is probably the seat of most metabolic activity.

Redia Intermediate larval stage of a digenean platyhelminth, produced in the mollusc intermediate host, that in turn produces cercariae.

Resilin The highly elastic protein that is responsible for the leap of the flea.

Rostellum Ring of hooks on the scolex of some cestodes.

Saefftigen's pouch Internal muscular sac at the posterior of male acanthocephalans, used to manipulate the copulatory bursa.

Sampling at autopsy The procedure of assessing the migration route of parasites by examining their position in the carcasses of hosts at various times after infection.

Schistosomulum Stage in the life-cycle of a schistosome between the invading cercaria and the adult.

Schizogony Process of asexual reproduction in sporozoan protozoans in which multiple mitoses are followed by cell divisions that result in large numbers of daughter cells at the same time.

Schizont Cell undergoing schizogony.

Scolex Anterior attachment organ of a cestode.

Scutum Sclerotised antero-dorsal plate found on hard ticks.

Seta (plural setae) Bristle or spine.

Sheath The cuticle of a nematode retained after incomplete moult.

Somatic migration Migration of some nematode larvae into the body tissues of the host where they normally become quiescent or trapped.

Spicules Posterior cuticular structures, normally paired, in male nematodes, used for orientation during mating.

Spiny proboscis Anterior attachment organ of acanthocephalans, often retractable.

Sporocyst Developmental stage usually within the oocyst of sporozoan protozoans. Alternatively the stage of a digenean platyhelminth that develops from the miracidium in the mollusc intermediate host.

Sporogony Multiple division and maturation of a sporozoan oocyst.

Sporozoa Group of parasitic protozoans with no obvious locomotory structures and possessing at some stage in their life-cycle an apical complex. Alternatively known as the Apicomplexa.

Sporozoite Cell resulting from sporogony.

Spring rise Increase in nematode egg production in sheep and cattle seen in the spring, also known as periparturient rise.

Stratum corneum The outer keratinised layer of the skin.

Striped layer Outer layer of the acanthocephalan tegument.

Strobila The ribbon of individual segments that constitutes the body of a cestode.

Stylet Anterior cuticular piercing organ used by plant parasitic nematodes to penetrate the cell walls of their hosts.

Sucker Attachment organ found on digeneans and the scolex of cyclophyllidean cestodes.

Symbiosis General term used to cover all intimate inter-specific associations.

Synoecious commensalism An inter-specific association in which one animal lives within the shell or body of another, also known as inquilism.

Tarsus The terminal segment of an insect's leg.

T-cell Lymphocyte that has been processed through the thymus gland, may be further divided into sub-sets including T-helper cells and cytotoxic T-cells.

Temephos Insecticide used against *Simulium* larvae in the Onchocerciasis Control Programme. Also known as Abate.

Temperature telescoping Process by which the time of development of free-living parasite stages decreases with rising temperature so that large numbers become available on the pasture at the same time.

Thrombus (plural thrombi) A blood clot that forms within a blood vessel.

Tibia The fourth of five segments of an insect's leg.

Tracheal migration Migration of nematode larvae within their hosts through the liver and lungs and thence back to the intestine *via* the tracheal ciliary escalator.

Transferrin A plasma protein that binds iron.

Transmammary infection Passage of parasites from mother to offspring in the milk.

Transovarian transmission Transmission of parasites from one generation to the next *via* the egg.

Transplacental infection Passage of parasites from mother to offspring directly across the placenta.

Transport host Inessential host in which a parasite can survive but does not develop, also known as a paratenic host.

Transtadial transmission Transmission of parasites from one stage of development to the next, normally across a moult.

Trematoda A grouping of platyhelminths including the Monogenea, Digenea, and Aspidogastrea. Now largely superseded but sometimes used as a synonym for the Digenea.

Trickle infection Process of infection in which the host is exposed to a small number of infective agents at regular intervals.

Trophozoite Active feeding stage of a protozoan as opposed to a cystic stage.

Tropical pulmonary eosinophilia A disease syndrome seen in some patients as a consequence of infection with *Wuchereria bancrofti*.

Tumour Necrosis Factor (TNF)   Two cytokines with cytotoxic effects which also have immunoregulatory effects.

Uniramous   Unbranched. Normally applied to arthropod appendages.

Uterine bell   Structure in a female acanthocephalan that sorts the eggs, allowing mature fertilised eggs to leave while the immature ones are passed back into the body cavity.

Variable surface glycoproteins   Glycoproteins on the surface of trypanosomes recognised by the host immune system.

Vector   Intermediate host, frequently an insect, that carries a parasite to its definitive host.

Villous atrophy   Reduction in size of the intestinal villi.

VIP latrine   Modified and improved pit latrine.

Virulence   Capacity of an organism to overcome the host's defences.

Visceral larva migrans   Disease syndrome in man caused by the migration of *Toxocara* spp. larvae.

Zoonosis   A disease transmissible from animals to man.

Index